THE
MARTIANS

NICK REDFERN

evidence of life on
the red planet

NEW
PAGE

This edition first published in 2020 by New Page Books,
an imprint of
Red Wheel/Weiser, LLC
With offices at:
65 Parker Street, Suite 7
Newburyport, MA 01950
www.redwheelweiser.com

ISBN: 978-1-63265-176-1
Library of Congress Cataloging-in-Publication Data available upon
request.

Cover design by Kathryn Sky-Peck
Cover photo by NASA
Interior by Jane Hagaman
Typeset in ITC New Baskerville and Lato

Printed in the United States of America
IBI

10 9 8 7 6 5 4 3 2 1

Contents

Introduction

Mars, our nearest planetary neighbor in the solar system, is an average of approximately 140 million miles from the Earth. Because the orbit of Mars is erratic, the distance will fluctuate during the Martian year. Not particularly big in size, it is dwarfed by all of the planets (Pluto is now classified as a dwarf planet) except Mercury. At the other end of the spectrum, however, Mars is home to the gigantic Olympus Mons, which is the biggest volcano in the entire solar system. The huge amount of iron oxide on Mars gives it a noticeable—famous, even—reddish color, hence its memorable nickname: the Red Planet. It has a blood-freezing average temperature of −81 degrees Farenheit, and its decidedly harsh atmosphere is comprised almost exclusively of carbon dioxide; aside, that is, from a small amount of water vapor.

There is no denying the fact that dusty, rocky, desert-like Mars is a very interesting world: For one thing, it has polar ice caps. And those ice caps are comprised of water. Yes, good old water; just like ours. Not only that, the amount of water on Mars is huge. And I do not exaggerate when I use the word *huge*. NASA's current estimates suggest that if all of the Martian water was melted, there would be enough to cover the entire planet up to a height of about 35 meters, which is undeniably amazing. The planet's days are strikingly similar to ours, in terms of length: Mars's days last for twenty-four hours and thirty-seven minutes. Mars's year, however, runs for 687 days.

For millennia, we, the human race, have been intrigued by Mars. Indeed, a good case can be made that we have an absolute affinity to the Red Planet—an affinity that dictated the development of early religions and even the world of conflict. As NASA notes:

> The Egyptians were the first to notice that the stars seemed "fixed" and that the sun moves relative to the stars. They also noticed five bright objects in the sky (Mercury, Mars, Venus, Jupiter, and Saturn) that seemed to move in a similar manner. They called Mars Har Decher—the Red One. . . . Greeks called the planet Ares after their god of war, while the Romans called it Mars. Its sign is thought to be the shield and sword of Mars.[1]

As for Mars, the god, N. S. Gill says: "Mars sired Romulus and Remus, making the Romans his children. He was usually called the son of Juno and Jupiter, just as Ares was taken to be the son of Hera and Zeus. The Romans named an area beyond the walls of their city for Mars, the *Campus Martius* 'Field of Mars.'"[2]

Mars was important to the Roman Empire when it came to the matter of something that we, as a species, are all too aware of: warfare. And sacrifice, too, as Patti Wigington states:

> Before going into battle, Roman soldiers often gathered at the temple of Mars Ultor (the avenger) on the Forum Augustus. The military also had a special training center dedicated to Mars, called the Campus Martius, where soldiers drilled and studied. Great horse-races were held at the Campus Martius, and after it was over, one of the horses of the winning team was sacrificed in Mars' honor. The head was removed, and became a coveted prize among the spectators.[3]

Then, there's the matter of the ties between Mars and the world of science fiction. Take, for example, H. G. Wells's classic sci-fi novel of 1897, *The War of the Worlds*. It's a story that sees the human race up against an invasion by hostile extraterrestrials that live on the Red Planet. Moving on, there are Edgar Rice Burroughs's classic fantasy tales that chronicle the life of one John Carter. He's a heroic, sword-wielding figure who has all kinds of adventures on faraway Mars, with the first story being published in 1912. Burroughs's novels proved to be incredibly popular with the public of that era. They included *A Princess on Mars; Swords of Mars; Synthetic Men on Mars; Thuvia, Maid of Mars, John Carter of Mars;* and . . . well . . . you surely get the picture.

On October 30, 1938, a radio-version of *The War of the Worlds* was broadcast on the Mercury Theater on the Air. It was directed by Hollywood legend Orson Welles. So realistic was the drama, it led at least some of the listeners to believe that the Martians really *were* invading. In 1967, Hammer Film Productions released *Quatermass and the Pit,* the title of which was changed for American audiences to *Five Million Years to Earth.* It's a gripping tale of ancient Martians who came to our world millions of years ago—and who genetically altered early, primitive humans. And, in a strange and very alternative way, the Martians of the movie are still wielding their extraordinary, almost-supernatural powers in 1960s-era London, England, deep inside the tunnels of the London Underground rail system.

In 1979, NBC broadcast a mini-series based on Ray Bradbury's science-fiction novel *The Martian Chronicles. Mission to Mars* was a big-bucks movie made in 2000 that starred Gary Sinise and Tim Robbins as astronauts who learn—to their amazement—that Mars is not the dead world that so many believe it to be. Then, in 2015, actor Matt Damon took on the role of astronaut Mark Watney in *The Martian,* a movie that saw Damon's character forced to find a way to survive in the harsh environment of Mars.

Moving away from the domain of sci-fi, there is the matter of the many and varied unmanned missions to Mars that have been undertaken by the United States, Russia, India, Japan, and the European Space Agency. The combined figure—of fly-bys, landings, and orbits of Mars—is close to fifty. In other words, the world's leading nations in the field of space flight and astronomy have a deep interest in Mars. As a species, it's almost as if we just cannot leave Mars alone, whether in mythology, history, science fiction, or ambitious missions to the planet itself.

And there is one other thing to be noted: Mars is an utterly dead world. At least, that's what we're told by NASA. No animal life, no plant life. Nothing. That is not the case, though. As you'll soon see, Mars is absolutely teeming with life, both flora and fauna, no less. Some of that life is highly advanced and incredibly old. There's no denying that the planet guards its incredible and disturbing secrets both diligently and carefully. It has done so for a long time—an amazingly long time. None of that, however, has

Mars, a world filled with extraterrestrial mysteries
and ancient secrets. (NASA)

stopped us from uncovering a wealth of what can only be described as "anomalies" on Mars—and on one of its two moons, Phobos. As will become apparent, and as our story develops and heads off in some very strange ways, those same anomalies will reveal the most incredible part of the story: the direct connection between Mars and the Earth, and between Martians and humans—both ancient and modern.

The Martians: Evidence of Life on the Red Planet is an in-depth study of the theory that Mars was once a world that teemed with life. Perhaps, even, life that is not at all too dissimilar to ours. Incredibly, the Martians may *still* be there. *Alive.* The questions that this book asks and answers include:

◊   Do Martians really exist?

◊   What did they look like?

◊   What kind of society did the Martians have?

◊   What was it that caused their world to become a harsh, desert-like environment?

◊   Did global-warming, asteroid strikes, or nuclear war ensure the extinction of the Martians? Could it have been all three?

◊   Is it possible that as the planet began to spiral into a state of environmental collapse, a number of Martians were able to flee the Red Planet and settle on Earth—and, perhaps, in doing so were inadvertently perceived as gods by early humans?

◊   Does Mars still retain remnants of its long-gone civilization?

◊   Are Martian artifacts strewn about the surface, just waiting to be found by the likes of NASA?

◊   Has NASA already found such evidence, but chosen to withhold such monumental finds from the public and the media?

◊   What is the truth of the so-called "Face on Mars," a massive structure that eerily resembles the famous Sphinx at Giza, Egypt?

◊ Incredibly, could the Martians still exist, deeply below the surface of the planet in secure installations that allow them to ensure their civilization continues?

◊ What do we know about the Martian environment, its atmosphere, and its landscape?

◊ What have the many missions to Mars undertaken by NASA and other agencies uncovered?

Those and many more questions will be addressed in depth.

CHAPTER 1

"The Human Race Has Something to Do with Mars"

IIIIIIIIIIIIIIIIIIIIIIIIIIIIIIIIIIIIII

Located on a specific area of Mars called Cydonia Mensae are two huge, battered, and bruised structures that, for decades, have intrigued and amazed the public, NASA scientists and engineers, the media, the CIA, and psychics. And that's just the tip of the iceberg. The most famous one of the two has become known as the Face on Mars. As for the other, it's referred to as the D&M Pyramid, for reasons that will very quickly become clear. And, the pair do not stand alone: There are multiple anomalies in that particular location. Collectively, they provoke imagery of a ruined, ancient city, of a once-thriving world, and of a civilization that is now long-extinct and that took its secrets with it when the end came in nightmarish, civilization-trashing fashion.

As for their sizes, the Face is approximately two kilometers in length, while the D&M Pyramid is more than one kilometer in height (almost two-thirds of a mile). In other words, we are talking about *gigantic* structures. That is, unless you believe that seeing something not too dissimilar to a visage on Mars is no different to seeing the face of Elvis in a cup of coffee. Or of Jesus in a bagel. On this very point of how eye-catching images might be interpreted—correctly vs. wildly—there's no doubt that the controversy provokes massive debate. It shows no signs of going away any time soon. Or later, even. Without doubt, it is the Face on Mars—more than any other oddity on the planet—that has driven the story that you are now reading and that continues to generate extreme controversy. So, how did we get from there to here, in terms of the development of the overall story?

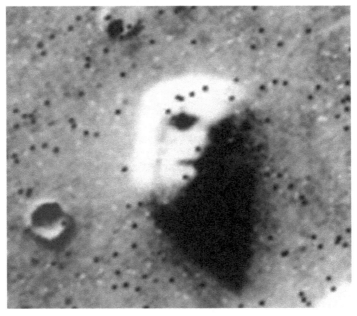

The enigmatic Face on Mars. (NASA)

In the years that have passed since the Face was first seen in the 1970s, matters have escalated to incredible degrees: Researchers claim to see a "fort" in the area of Cydonia, maybe even the remains of ancient roads. Today, the claims, speculations, and conclusions have reached absolute fever-pitch level. An intelligently crafted monolith; strange, crab-like creatures roaming the Martian land-scape; an ancient, humanoid-like skull; and even the image of the Egyptian queen, Nefertiti, have all been seen on Mars—or are natural phenomena that have been grossly misinterpreted. Before we address all of these issues, and many more, we need to go back in time to 1976, the year in which the mystery began. To acquaint you with the Face and its attendant controversies I will share with you the work of one of the most respected researchers of the Martian mysteries—something that will give you insight into how the puzzle began and why and how it's still very much with us.

As I noted, the Face is actually just one of a handful of unusual fixtures on a certain area of Mars called Cydonia, which is specifically within Cydonia Mensae. A portion of it is on the fringes of the Arabia Terra region and the Acidalia Planitia plain. Highlands and plains abound and dominate the Martian landscape. Something else dominates the landscape, too. That's right: that enigmatic face-like structure. Or, a perfectly natural, huge mesa, depending on your very own perspective.

While numerous researchers have penned papers and books on the matter of the Face on Mars, for me the most significant of all was the late Mac Tonnies. And why, you may ask, was Tonnies's work so important and vital to the investigations? And why is it *still* so important? Simply put, Tonnies was neither a full-on believer nor a skeptic when it came to the massive Cydonia structures. He was someone who was just looking for the answers, regardless of where they might have taken him. Tragically, Tonnies died in 2009, at the age of just thirty-four, from the effects of a heart condition. He left an incredible body of data behind, however, the vast majority of it is presented in the pages of his 2004 book, *After the Martian Apocalypse.*

I was lucky enough to interview Tonnies before he passed away, specifically on the salient points of the overall controversy—something that will quickly acquaint you with the much wider and broader story that this book tells. Tonnies shared with me his thoughts and conclusions on the entire controversy surrounding the Face, something that makes it clear that he strongly suspected that at least *some* of the curious structures on Mars were manmade (or, to be entirely accurate, Martian-made), even if he didn't buy into everything that caught the attention of the eyes and ears of other researchers of the Face on Mars. Tonnies had so much to say, in what turned out to be such a little amount of time, but he was careful to pinpoint the primary aspects of the debate concerning what was—or what wasn't—on Mars: massive structures constructed by intelligent beings from a faraway world.

Most important of all, Tonnies's words perfectly set the scene for what is to come in the pages ahead.

In our Q&A,[1] Tonnies began as follows:

I've always had an innate interest in the prospect of extraterrestrial life. When I realized that there was an actual scientific inquiry regarding the Face and associated formations, I realized that this was a potential chance to lift SETI [the Search for Extraterrestrial Intelligence] from the theoretical arena; it's within our ability to visit Mars in person. This was incredibly exciting, and it inspired an interest in Mars itself—its geological history, climate, et cetera. I have a BA in creative writing. So, of course, there are those who will happily disregard my book because I'm not "qualified." I suppose my question is "Who *is* qualified to address potential extraterrestrial artifacts?" Certainly not NASA's Jet Propulsion Laboratory, whose Mars exploration timetable is entirely geology-driven.

As for how NASA learned of the Face on Mars—something that directly led the rest of us to hear of it, and to finally see it, too—Tonnies laid down the facts:

NASA itself discovered the Face, on July 25, 1976, and even showed it at a press conference, after it had been photographed by NASA's *Viking* mission probe. Of course, it was written off as a curiosity. Scientific analysis would have to await independent researchers. The first two objects to attract attention were the Face and what has become known as the "D&M Pyramid." Both of them were unearthed by digital imaging specialists Vincent DiPietro and Gregory Molenaar [at the time, engineers at NASA's Goddard Space Center at Greenbelt, Maryland]. Shortly after, Face researcher Richard Hoagland pointed out a collection of features near the Face which he termed the "City." The "Fort," too.

On the matter of the Fort, Tonnies sent me the following in our email-based interview: "The Fort looks weathered, defeated. Its eastern side is riddled with small, shallow craters that terminate as abruptly as the holes left from a burst of machine gun-fire. Indeed, it's easy to imagine that you're examining the sterile ruins of some unimaginable conflict."

On that point of an "unimaginable conflict," Tonnies might have been right on target, as we shall later see. The City Pyramid has a "peculiar fives-sided shape," he said. And, then, there's the D&M Pyramid. Tonnies put it like this: "The D&M's surface is not the smooth finish found elsewhere in Cydonia. Rather, its shallow incline is swollen and cracked, as if once molten. Despite this, no signs of volcanism are apparent." Tonnies also noted: "Interestingly, there seems to be a tunnel-like opening into the D&M. If the D&M is an extraterrestrial structure, then perhaps you'll find evidence there proving beyond doubt that civilized Martians once existed."

Under what circumstances did the debate concerning Cydonia begin? Tonnies told the story:

When NASA dismissed the Face as a "trick of light," they cited a second, disconfirming photo allegedly taken at a different sun-angle. This photo never existed. DiPietro and Molenaar had to dig through NASA archives to find a second image of the Face. And,

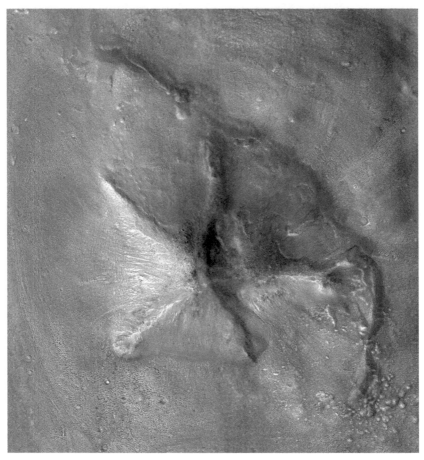

Mars's legendary D&M Pyramid, pristine no more. (NASA)

far from disputing the face-like appearance, it strengthened the
argument that the Face remained face-like from multiple viewing
angles.

Tonnies then turned his attentions even more in the direction
of NASA:

The prevailing alternative to NASA's geological explanation—that
the Face and other formations are natural landforms—is that we're

seeing extremely ancient artificial structures built by an unknown civilization. NASA chooses to ignore that there *is* a controversy, or at least a controversy in the scientific sense. Since making the Face public in the 1970s, NASA has made vague allusions to humans' ability to "see faces" (e.g., the "Man in the Moon") and has made lofty dismissals, but it has yet to launch any sort of methodical study of the objects under investigation. Collectively, NASA frowns on the whole endeavor. Mainstream SETI [the Search for Extraterrestrial Intelligence] theorists are equally hostile.

Tonnies made an important observation:

Basically, the Face—if artificial—doesn't fall into academically palatable models of how extraterrestrial intelligence will reveal itself, if it is in fact "out there." Searching for radio signals is all well and good, but scanning the surface of a neighboring planet for signs of prior occupation is met with a very carefully cultivated institutionalized scorn. And, of course, it doesn't help that some of the proponents of the Face have indulged in more than a little baseless investigation.

Now, let's see what Tonnies thought was the "real truth" of the debate: Addressing this issue over the phone[2], he said to me: "I *think* some of the objects in the Cydonia region of Mars are probably artificial. And I *think* the only way this controversy will end is to send a manned mission. The features under investigation are extremely old and warrant on-site archaeological analysis. We've learned—painfully—that images from orbiting satellites won't answer the fundamental questions raised by the Artificiality Hypothesis."

I asked Tonnies: "Do you believe all the perceived anomalous structures are indeed that or do you feel some are of natural origin while some are of unnatural origin?" His reply: "I suspect that we're seeing a fusion of natural geology and mega-scale engineering. For example, the Face is likely a modified natural mesa, not entirely

Cydonia: Martian ruins from the distant past. (NASA)

unlike some rock sculptures on Earth but on a vastly larger and more technically challenging scale."

All of the talk of pyramids—and, later, of opinions that the Face had a Sphinx-like appearance—led me to pose questions to Tonnies that I wasn't at all sure he would be willing to answer, lest he might be accused of over-exaggerating and sensationalizing the situation. My questions were these: Is there a relationship between the face and the pyramids on Mars and the similar ones at Giza, Egypt? What does the research community think of this perceived connection?

Tonnies, I was pleased to see, *was* willing to tackle the questions:

There's a superficial similarity between some of the alleged pyramids in the vicinity of the Face and the better-known ones here on Earth. This has become the stuff of endless arcane theorizing, and I agree with esoteric researchers that some sort of link between intelligence on Mars and Earth deserves to be taken seriously. But, the formations on Mars are much, much larger than terrestrial architecture. This suggests a significantly different purpose, assuming they're intelligently designed. Richard Hoagland, to my knowledge, was the first to propose that the features in Cydonia might be "arcologies"—architectural ecologies—built to house a civilization that might have retreated underground for environmental reasons.

This Mars–Egypt issue will surface time and time again as this book progresses, and as we dig further into this particular controversy.

I had another question for Tonnies: If these things are artificial, who built them: Martians, someone visiting Mars, ancient Earth civilizations now forgotten or lost to history?

Tonnies answered:

It's just possible that the complex in Cydonia—and potential edifices elsewhere on Mars—were constructed by indigenous Martians. Mars was once extremely Earth-like. We know it had liquid water. It's perfectly conceivable that a civilization arose on Mars and managed to build structures within our ability to investigate. Or, the anomalies might be evidence of interstellar visitation—perhaps the remains of a colony of some sort. But why a humanoid face?

That's the disquieting aspect of the whole inquiry; it suggests that the human race has something to do with Mars, that our history is woefully incomplete, that our understanding of biology and evolution might be in store for a violent upheaval. In retrospect, I regret not spending more time in the book addressing the possibility that the Face was built by a vanished terrestrial civilization that had achieved spaceflight. That was a tough notion to swallow, even as speculation, as it raises as many questions as it answers.

If the Cydonia "items" were constructed and not wholly natural formations, then when might all of this have gone down? Tonnies's answer was a swift and short one: "We need to bring archaeological tools to bear on this enigma. When that is done, we can begin reconstructing Martian history. Until we visit in person, all we can do is take better pictures and continue to speculate."

As for how an allegedly once-thriving world was turned into a largely dead one, Tonnies had his thoughts:

> Astronomer Tom Van Flandern has proposed that Mars was once the moon of a tenth planet that literally exploded in the distant past. If so, then the explosion would have had severe effects on Mars, probably rendering it uninhabitable. That's one rather apocalyptic scenario. Another is that Mars' atmosphere was destroyed by the impact that produced the immense Hellas Basin [a 7,152-meter-deep basin located in Mars' southern hemisphere]. Both ideas are fairly heretical by current standards; mainstream planetary science is much more comfortable with Mars dying a slow, prolonged death. Pyrotechnic collisions simply aren't intellectually fashionable—despite evidence that such things are much more commonplace than we'd prefer.

With the questions concerning Mars now answered, I turned my questions to Tonnies himself: What was it that prompted him to write *After the Martian Apocalypse*?

This was Tonnies's reply:

> Anger. I was, frankly, fed up with bringing the subject of the Face on Mars up in online discussion and finding myself transformed into a straw man for self-professed experts. It was ludicrous. The book is a thought experiment, a mosaic of questions. We don't have all of the answers, but the answers are within our reach. Frustratingly, this has become very much an "us vs. them" issue, and I blame both sides. The debunkers have ignored solid research that would undermine their assessment, and believers are typically quite pompous that NASA et al are simply wrong or, worse, actively covering up.

So, did Tonnies think that NASA was engaged in a cover-up to hide the truth, regardless of what it might be? Yes, he did—but most certainly not to the extent that certain conspiracy theorists have suggested in more recent years, as you will later learn. Tonnies spelled it out for me:

When NASA/JPL [Jet Propulsion Laboratory] released the first *Mars Global Surveyor* image of the Face in 1998, they chose to subject the image to a high-pass filter that made the Face look hopelessly vague. This was almost certainly done as a deliberate attempt to nullify public interest in a feature that the space agency is determined to ignore. So yes, there *is* a cover-up of sorts. But it's in plain view for anyone who cares to look into the matter objectively. I could speculate endlessly on the forms a more nefarious cover-up might take, but the fact remains that the *Surveyor* continues to return high-resolution images. Speculation and even some healthy paranoia are useful tools. But we need to stay within the bounds of verifiable fact lest we become the very conspiracy-mongering caricatures painted by the mainstream media.

The Q&A came to its end with this: "What do you hope your book *After the Martian Apocalypse* will achieve?"

Tonnies reply came as follows:

Our attitudes toward the form extraterrestrial intelligence will take are painfully narrow. This is exciting intellectual territory, and too many of us have allowed ourselves to be told what to expect by an academically palatable elite. I find this massively frustrating. I hope *After the Martian Apocalypse* will loosen the conceptual restraints that have blinkered radio-based SETI by showing that the Face on Mars is more than collective delusion or wishful thinking. This is a perfectly valid scientific inquiry and demands to be treated as such.

As we will now see, and as further Martian enigmas are constantly being found, years after Tonnies's book was published the whole controversy still firmly remains as "a perfectly valid scientific inquiry."

There's another aspect to all of this, too: namely that, as the "Face-based" research began to expand—almost exponentially, it seemed, at times—evidence began to surface suggesting that the mysteries of Mars did not actually begin in the 1970s, at all. Incredibly, they went back centuries. The Face on Mars, which you are now acquainted with, is just a part of the story.

Let's now address the rest of it. It's time to take a deep look at the overall controversy and see where this story began and where it ultimately takes us now. You may be surprised at what you are about to learn: Ancient secrets don't always stay that way. In this particular case, they hit us square in the face (no pun intended) and provoke us to contemplate and wonder on the stunning possibility that Mars was once a world not at all unlike ours—and a world filled with beings not unlike us, too.

CHAPTER 2

"Two Lesser Stars, or Satellites, which Revolve about Mars"

||||||||||||||||||||||||||||||||||||||||

I t's now time for us to turn our attentions to an intriguing mystery that offers the possibility that, centuries ago, a famous author had secret access to ancient information surrounding the planet Mars. It's a mystery that revolves around none other than Jonathan Swift, the brains behind the classic novel *Gulliver's Travels*. The book was first published in 1726, almost three hundred years ago. Yet, as we shall soon see, there are good reasons for concluding that Swift had in his possession arcane knowledge of the Red Planet that extended much further into the past than the eighteenth century. Before we get to the matter of Swift's strange connection to Mars, though, let's first take a look at the storyline of his famous novel. The planned, original title of *Gulliver's Travels* was *Travel into Several Remote Nations of the World in Four Parts . . . by Lemuel Gulliver.*[1] It's hardly surprising that with a title like that the publisher chose to trim it down a bit. The story tells of the adventures of the aforementioned Lemuel Gulliver. We learn that he has a long affinity for the vast oceans of our world, that he is a skilled surgeon, and that he has traveled the planet, overseeing the crews of several ships. Swift's story sees our hero head out to a series of tongue-twisting places, including Lugg-nagg, Laputa, and Brobdingnag, where adventures abound.

Although Swift's acclaimed novel was first published in 1726, the man himself had been working on the story as far back as 1713. It wasn't until a full seven years later that Swift finally decided to seriously get into the writing of what was, until then, a largely stalled manuscript. Swift certainly made up for lost time, however, toiling night and day to get the book whipped into good shape. He certainly did that. On reading Swift's work, Benjamin Motte, a London publisher, eagerly decided to publish it. It was a most wise move: *Gulliver's Travels* became a distinct bestseller. It's still a crowd-puller to this day: In 2010, a big-bucks movie version starring Jack Black reaped in more than $200 million. Now, we come to the matter—and the mystery—of Swift's connection to Mars.

In *Gulliver's Travels,* we learn that the people of Laputa, an island that Gulliver visits during the course of his worldwide trek, have ". . . discovered two lesser stars, or satellites, which revolve about Mars, whereof the innermost is distant from the center of the primary planet exactly three of its diameters, and the outermost five; the former revolves in the space of ten hours, and the latter in twenty-one and a half; so that the squares of their periodical times are very near the same proportion with the cubes of their distance from the center of Mars, which evidently shows them to be governed by the same law of gravitation that influences the other heavenly bodies."[2]

One might say that there is nothing particularly strange about all of this. After all, Mars does indeed have a pair of small satellites. Their names are Deimos and Phobos. So, why is there such a controversy surrounding Swift's words? The answer is a fascinating and tantalizing one: Mars's two moons were not discovered until 1877—approximately a century and a half *after* Swift's book was published. This begs an important and glaringly obvious question: How could Swift have known of the existence of the Martian moons long before anyone else? Certainly, in the 1700s there were no telescopes that could secure a viewing of the two moons. Not only that, Swift's descriptions of Deimos and Phobos were not too dissimilar to the reality of the situation.

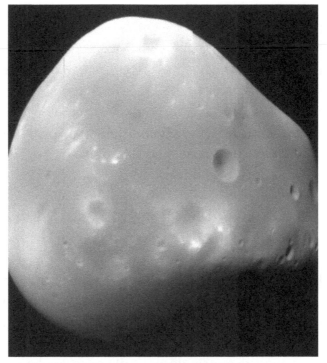

Deimos's big mystery: the secret world of Jonathan Swift. (NASA)

On this issue, astronomer David Darling says:

When the two Martian moons, Phobos and Deimos, were eventually found, by Asaph Hall at the US Naval Observatory, their orbits proved to be quite similar to those described in Swift's novel. Phobos is actually 6,000 km from the surface of Mars and revolves around

Mars in 7.7 hours, whereas Swift gave the values 13,600 km and 10 hours, respectively. Deimos averages 20,100 kilometers from Mars and orbits in 30.3 hours; Swift gives 27,200 kilometers and 21.5 hours, respectively.[3]

Adding even more intrigue to the story: Swift was not the only writer, in the 1700s, to address the matter of Mars's two moons. They pop up in the pages of an early sci-fi story, *Micromegas*, written in 1750 by Voltaire, an acclaimed French philosopher and historian. Of course, the fact that Voiltaire's story came *after* Swift's makes it not at all impossible that Voltaire simply decided to take a degree of inspiration from the pages of *Gulliver's Travels*. There is another possibility, too: It is focused on what is termed as *celestial harmony*. It's a controversial and now-dismissed theory that was postulated by a German astrologer and astronomer, Johannes Kepler, who was born in 1571 and who died in 1630. Kepler's theory was that the numbers of moons that orbited the planets in our solar system adhered to a certain concept: that of celestial harmony. Kepler, as a revered astronomer, concluded that neither Venus nor Mercury had moons. He was correct. Obviously, he knew that our planet has just one moon. And, in the world of celestial harmony, that meant Mars would likely have two moons, Jupiter would have three, and Saturn four. And so on and so on.

Today, we know that the numbers of moons that exist in our solar system most definitely do not adhere to Kepler's concept. It was a good try though. So, that means both Kepler and Swift had gotten the numbers of moons for Mars correct, but that it was all down to chance and coincidences. Or was it? Let's not forget the words of David Darling, who noted that, when discovered, the orbits of Deimos and Phobos " . . . proved to be quite similar to those described in Swift's novel."[4]

In light of Darling's words, it's time to look at an even more incredible theory.

Phobos: a Martian moon and hidden, arcane knowledge. (NASA)

Was Kepler's concept of celestial harmony the cause of all the controversy surrounding the Martian moons? Was the whole matter a bizarre coincidence? Or, did Jonathan Swift, as a Freemason, have access to ancient secrets pertaining to a now-long-forgotten Martian race? They are questions that offer us a great deal of food for thought. So far, however, the questions remain unanswered. Right now, we're working with what is largely rank speculation—but, it's most certainly fascinating speculation.

CHAPTER 3

"This Is a Message from Another Planet, Probably Mars"

||||||||||||||||||||||||||||||||||||

As I noted in Chapter 1, Mac Tonnies speculated on the possibility that there may have been a connection between an ancient Martian race and the world of equally ancient Egypt. The Face on Mars, the D&M Pyramid, and the Nefertiti-like imagery (the latter of which we'll come to investigate later) have all been championed as evidence of a link between our world and that of the Red Planet. While a case can certainly be made that the connection is a real one, it's important to note that there is nothing new about the Martian-Egyptian issue. The mainstream media was reporting on such a particular connection way back in the late 1800s. I should stress that this does not necessarily mean that the nineteenth-century tales had a direct role in influencing today's researchers of this highly controversial Mars–Egypt matter. On the other hand, we should not wholly write off such a theory as a possibility.

It's intriguing to note the time frame of this particularly odd story: the late 1890s. This was the period in which the United States was hit by a veritable wave of UFO activity, specifically in the years of 1896 and 1897. HowStuffWorks says of this curious flap of aerial activity:

> Between the fall of 1896 and the spring of 1897 people began sighting "airships," first in California and then across most of the rest of the United States. Most people (though not all) thought the airships were machines built by secret inventors who would soon dazzle the world with a public announcement of a break-through in aviation technology leading to a heavier-than-air flying machine. More than a few hoaxers and sensation-seeking journalists were all too happy to play on this popular expectation. Newspaper stories quoted "witnesses" who claimed to have seen the airships land and to have communicated with the pilots.[1]

To be sure, it was a wild, crazy, and unpredictable time, one that came decades before the terms "UFOs" and "Flying Saucers" were on the minds of anyone. That period was made even more controversial by the fact that a Martian-Egyptian story caught the attention of the media of the day. The story is a winding and twisting one, a saga that became more and more entangled as time went on.

Jason Colavito, a noted skeptic when it comes to matters relative to the "ancient aliens" phenomenon, gives us a solid introduction to this story:

> Throughout 1897 American newspapers frequently reprinted an undated and vague story about a Belgian man who claimed to have been injured by a falling meteorite containing hieroglyphics. Later that year a physician and astrologer in Binghamton, New York named James MacDonald claimed to have found a similar meteor. This produced a bit of a media frenzy as quack earthquake-predicting scientist E. S. Wiggins agreed with MacDonald that the object contained writing from Martians."[2]

Now, let's take a look at the bigger picture.

So far as can be ascertained, on November 14, 1897, the story caught the attention of the American public on a large scale. That was the date on which none other than the *New York Times* splashed the tale across its pages. It all began with the curious affair of something crashing to earth in Binghamton, New York. According to Professor Jeremiah MacDonald, then of Park Avenue, he was heading home late at night when his attention was caught by a sudden, bright flash, which coincided with something slamming into the ground—to the extent that it was buried into the soil. The *New York Times* reported that after the whatever-it-was was extracted from the ground, it was seen to be a "mass of some foreign substance that had been fused by intense heat."[3]

As the excavation occurred, the object was found to still be emitting a huge amount of heat. When the thing from the skies had finally cooled down, and with the help of plenty of cold water, it was broken wide open. The air was filled with a nauseating odor, not unlike that of sulfur. It is at this point that we see the first reference to what, in an odd fashion, amounted to a Mars–Egypt link, as the *New York Times* revealed: "Inside was found what might have been a piece of metal, on which were *a number of curious marks that some think to be characters* [author's emphasis]."[4] Those "characters" would very soon be perceived as something akin to Egyptian hieroglyphics.

The *Times* concluded in its article:

Prof. Whitney of the High School declared it an aerolite, but different from anything he had ever seen. The metal had been fused to a whitish substance, and is of unknown quality to the scientific men who have examined it. Several have advanced the opinion that this is a message from another planet, probably Mars. The marks bear some resemblance to Egyptian writing in the minds of some. Prof. McDonald is among those who believe the mysterious ball was meant as a means of communication from another world.[5]

And thus was cemented an almost symbiotic link between the Red Planet and Egypt. It should be noted that the *Times* was very careful as to how it worded matters relative to the story. Its journalists were not saying that the Egyptian empire of millennia ago was somehow interlinked with Mars. Only that the object demonstrated "some resemblance" to Egyptian hieroglyphics. For the public though, amazed by the story, the tantalizing threads were all but there to see. Four days later, the *New York Times* reported yet further on this controversy-filled matter: "Prof. [Ezekiel Stone] Wiggins believes that the aerolite that fell near Binghamton a few nights ago, and is alleged to have contained a piece of iron with hieroglyphics, was really a message from Mars."[6]

As for the professor himself, he said:

My opinion is that stones have for many thousands of years fallen from space upon the earth which actually contained written characters. The ancient Jews and other nations speak of their sacred books as having fallen from heaven. As the earliest important records were preserved in stone, it seems probable that the idea originated with aerolites, like that of Binghamton. There is no doubt in my mind that there are thousands of these stones that have fallen to our planet since man arrived here and are messages from another planet.[7]

Purely as an aside, it should be noted that Wiggins had quite a personal obsession with Mars: When an outbreak of Yellow Fever occurred in Jacksonville, Florida, in 1888, Wiggins told the press:

The planets were in the same line as the sun and earth and this produced, besides Cyclones, Earthquakes, etc., a denser atmosphere holding more carbon and creating Microbes. Mars had an uncommonly dense atmosphere, but its inhabitants were probably protected from the fever by their newly discovered canals, which were perhaps made to absorb carbon and prevent the disease.[8]

As Wiggins's words demonstrate, he was not just a believer in life on Mars—intelligent life, it's important to stress—but he had also bought into the "Martian canals" controversy, too, something that further expanded the theory of Mars being an extraordinary world, and possibly one with a highly advanced culture. We'll get to those canals soon. All of this was eagerly absorbed by the excited public. It should be noted, too, that Wiggins also had a fascination with unknown animals, such as lake monsters and sea serpents. It's a field known as cryptozoology, in which, today, research into the likes of Bigfoot, the Chupacabra, and the Abominable Snowman dominates.

The press had an important question for Wiggins: How did the Martian artifact reach Earth? Wiggins had an answer that was thought-provoking, but clearly speculative in nature:

> If we lived on Mars and possessed the scientific knowledge of those people, we might accomplish the feat without difficulty—in fact in a few hundred years more we may be able to generate and so control electric force that we can throw a projectile beyond the moon's orbit, so that would either fall satellite or move toward and fall upon another planet. Suppose the Martians were to throw a stone highly electrified into the orbit of their nearest satellite, which is only about 7,000 miles away, so that it would be in advance of it in its orbital motion. I have no doubt it would repel the stone in a line of a tangent and with such force as to send it to our planet's orbit, or, suppose a comet were passing near Mars and toward the earth, stones thrown near it would follow in its trail and fall to the earth like the stones which fell to the earth in November, 1872, after the comet of that year had crossed our planet's orbit.[9]

On November 21, 1897, the *New York World* got involved in the growing story. The headline of their article was guaranteed to make the readers of the newspaper sit up and take close and careful notice: "Is This Meteor a Message from Faraway Mars?" It was a loaded question, without any shadow of a doubt. In relation to its very own question, the *New York World* said: "The people who live on the planet of Mars may have been making efforts, from time to time through the centuries, to establish talking relations with the people who live on this planet. Some of the most thoughtful astronomers and scientists have declared it to be possible."[10]

The article continued:

> M. Flammarion, the famous French astronomer, has pointed out some things that are seen through the telescope on the surface of our planetary neighbor which justify the supposition that there are people living there, and that they have been trying to attract our attention by a system of geometric signals. He has observed certain luminous points on the surface of Mars which appear to have been placed very regularly, as if they were intended to mean something.

It was stressed, however, that

> [t]he signals would have to be on an enormous scale. A huge triangle traced in luminous lines on the lunar surface, each side of the triangle being perhaps twelve kilometers long, would be visible here on the earth with the aid of telescopes.
>
> Some astronomers have, therefore, inferred that if such triangles, or circles, or squares, were constructed on the surface of one of the large plains of the earth, made luminous by day by solar rays and at night by electricity, people on the Moon, if there are any, could see them, provided they have telescopes as good as we have. Seeing them, it follows that they could answer them with other luminous geometric signals. Soon afterward, we should be able to talk with them, and exchange all our information for theirs.

So, in the contents of this particular article, we have yet *another* aspect of the Mars controversy that intrigues so many Mars researchers today: that of huge images seen from the heavens above. One could say that triangles, circles, and squares are not too different to a massive face on Mars, at least in terms of what their purpose might have been.

The following, from *The Academy,* on December 11, 1897, repeats some of the data provided earlier in this chapter. It is instructive, however, as it brings H. G. Wells's classic novel *The War of the Worlds*—a book that told of an invasion of the Earth by hostile Martians—into the fray. On December 11, 1897, *The Academy's* readers were treated to these words:

> Fiction is continually giving Nature hints, of which she avails herself. No sooner is Mr. Wells's Martian story, *The War of the Worlds*, finished, than the report reaches us from America of an aerolite which has been found at Binghamton, New York. . . . The theory is, that the written characters form a message addressed to us from another world, probably Mars. We regret that the projectile fell in a land so prodigal of tall tales as America, but we congratulate Mr. Wells.[11]

The final account of any significant importance appeared in the *Windsor Magazine* in 1905, long after the story had originally surfaced. The article was titled "Things That Fall from the Sky" and was penned by one Walter George Bell. The article does not place the matter in a particularly good light:

> Some time ago we were excited by news that a message had come to Earth from Mars. It took the form of an aerolite, and dropped conveniently near the garden of Professor Jeremiah McDonald, at Binghampton, New York State. The professor was making his way home in the early morning hours, when, in a blinding flash of light, an object buried itself in the ground near him.
>
> On being dug out it proved to be a metallic mass which had been fused by intense heat. When cooled and broken open, we were told, "inside was found what might have been a piece of metal, on which were a number of curious marks like written characters"—which

characters, it was interesting to learn, "bore some resemblance to Egyptian handwriting."

Mars is our neighboring world. A popular belief has grown up in the existence of intelligent beings on Mars. So here, indeed, was a message from Mars! A delightful story, certainly; but attempts to read this "message" can only be so much time wasted. It was the "metal inside" which racked the brains of the Yankee reporter, and suggested to him that the message had been wrapped by careful Martians in a casing of another metal, black in color; but both are one and the same. A black casing, or rind, is common to all aerolites, and is created by fusion of the surface by the intense heat set up by friction with our atmosphere, as the aerolite dashes through to earth. As to the "message" in unreadable hieroglyphics, figures of the kind are not uncommon, and are largely due to crystallization.[12]

Jason Colavito suggested that a hoax on the part of MacDonald may have been the cause of the whole matter. You, after digesting the above words, may disagree and may conclude that something genuinely intriguing came down on that fateful, dark night way back in 1897. Today, as a result of the passage of more than a century, we'll never know for sure what happened. Whatever the answer, the fact is that more than a century ago the American public, in particular, was exposed to a fascinating—albeit sensational—story of Mars, of Egypt, of "Egyptian and Hebraic signs," of canals that actually weren't canals, and of huge markings on the ground that were—in Face on Mars–style—designed to be seen from above. In light of all of this, one has to wonder if any of this nineteenth-century story, and its attendant spin-offs, had a bearing and an influence on the minds of at least *some* of those who seek remarkably similar, strange phenomena on the surface of Mars today. And who seek a Mars–Egypt connection.

Now, let's take a much closer look at those canals that had the media intrigued and also how the phenomenon has taken on a significant role in today's world of Martian mysteries.

CHAPTER 4

"Mars as the Abode of Life"

||||||||||||||||||||||||||||||||||||||||

Imagine, if you will, a huge network of strange, huge "tubes" positioned across certain portions of Mars's vast landscape. Not only that, they just might have been the sophisticated creations of ancient Martians who used them as a means of near-planet-wide transport in the distant past. Sounds incredible? Well, yes, it *is* incredible. But does that make it true? That's the big question that sorely requires answering. The controversy began after certain photos of the curious structures were secured by NASA's *Mars Global Surveyor (MGS)*, which was launched in November 1996. The *MGS* obtained incredible images from all across the planet. Indeed, as NASA noted:

> *Mars Global Surveyor,* launched in 1996, operated longer at Mars than any other spacecraft in history, and for more than four times as long as the prime mission originally planned. The spacecraft returned detailed information that has overhauled understanding about Mars. Major findings include dramatic evidence that water still flows in short bursts down hillside gullies, and identification of deposits of water-related minerals leading to selection of a Mars rover landing site.[1]

For those who believed (and those who continue to believe) that the *MGS* photographed much more than the planet itself—namely, incredible, intelligently crafted megastructures on Mars—this was evidence that, even if the Martians were long gone, some portion of their civilization, at least, still existed. *Somewhere.* And, incredibly, could still be seen. It was a truly exciting time—and a new development, too—for those who championed the belief that the Face on Mars and the D&M Pyramid were the work of extraterrestrials who were born, lived, and died on the Red Planet in the distant past. Before we get to the matter of "Martian tubes" ("glass tubes," as they variously became known), it's very important to note that more than a century earlier, scientists and astronomers got all hot and bothered by what they thought were nothing less than canals on Mars. The parallels between what was found (or, rather, what was *assumed* to have been found) in the nineteenth century and those mysterious tubes discovered in the final years of the twentieth century are plain to see—and they demonstrate that before we get too excited it's vital we keep a collective, balanced head on our equally collective shoulders.

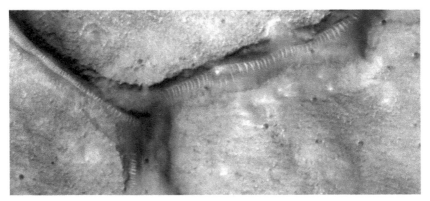

Mars's enigmatic "glass tubes." (NASA)

More than anything else, it's important to understand that the nineteenth century was the period in which astronomers, armed with more and more sophisticated telescopes, began to make astonishing discoveries on the surface of Mars. One of those astronomers was Giovanni Virginio Schiaparelli. He ultimately rose to the position of director of the Brera Astronomical Observatory, which is located in Milan, Italy, and was constructed in 1764, the work having been overseen by astronomer Roger Boscovich. Schiaparelli was enthused to the max by the prospect of learning more and more about Mars. While Schiaparelli was a well-respected astronomer, he was also someone who reached just a little too far when it came to the matter of trying to understand what, exactly, was happening on

The controversy of the canals. (Wikimedia Commons)

Mars. Certain light and dark portions of Mars were interpreted by Schiaparelli as oceans and what he even termed as "continents"— terminology that understandably suggested that Mars was not too dissimilar to our Earth. How very wrong he turned out to be. There is also the matter of the canals of Mars that excited the astronomer.

During the course of his studies of Mars, Schiaparelli's work led to a most catastrophic blunder; it was a blunder that would never, ever be forgotten. Granted, it wasn't entirely Schiaparelli's fault, but he certainly played a significant role in the development of the embarrassing matter. While scrutinizing Mars with his telescopes, Schiaparelli was sure that he could see unusual "channels" on the surface of the planet. He went on to refer to them as *canali*. When the word got out to an excited media, that word—*canali*—was one hundred percent erroneously translated to *canal*. It's very important to note the time frame: This was a particular era in human civilization when canals were rightly seen as vital components of then-modern-day society. As an example, in 1869 the 120-mile-long Suez Canal was completed and connected the Red Sea and the Mediterranean Sea. It was a major development of engineering. Quite understandably, for many people of that era, then, the idea of canals on Mars near-instantaneously suggested they were the work and the creation of intelligent beings. After all, that we were constructing vast canals at the very same time, added more and more weight to the theory that Martians might be doing precisely likewise. Thus was born the myth of the Martian canals. It was a myth that didn't go away.

Another astronomer who bought into the canal theory was Percival Lowell. He wasn't just an astronomer, but a skilled businessman, too—something that ensured him a sizeable fortune. Born and bred in Boston, Massachusetts, Lowell was the brains behind the Flagstaff, Arizona–based Lowell Observatory. His book *The Evolution of Worlds* was published in 1910. He had a deep affinity with Japan, having spent a great deal of time there and getting to know the people and their culture. And, yet, for all of his skills as an astronomer, Lowell also bought into the theory that there were canals on Mars—and not just canals made naturally, but canals created by intelligent Martians. Such was Lowell's ever-growing fascination with Mars, he wrote three books on the planet: 1895's *Mars; Mars and its Canals,* which was published in 1906; and *Mars as the Abode of Life,* which appeared two years later. The further that Lowell delved deep into his theory, so a larger, and far more elaborate, scenario came to the forefront. He personally weaved a novel and admittedly engaging picture that involved the Martians using huge numbers of canals to move massive amounts of water from the polar regions to Mars's equator. The whole thing was carefully, and even ingeniously, worked out. Unfortunately for Lowell, he was wrong. *Dead wrong.* History, science, and the ever-growing increase in sophisticated telescopes and probes have since made it very clear that there are no intelligently made canals on Mars.

Anywhere. What was once seen as a major development in the quest to find Martians is now nothing but an embarrassing affair that is best forgotten. Forever.

Does that also mean we should dismiss tales and theories of intelligently made huge tubes on the Red Planet today? Are the mysterious tubes of present-day time nothing more than the modern-day, real equivalents of the nonexistent canals that captivated both Lowell and Schiaparelli a long, long time ago? Was this really solid evidence, caught on camera by NASA, of incredible structures created by advanced Martians? Or, was a huge wave of embarrassment about to engulf those whose enthusiasm just might cause them to jump the gun? Those were the questions being asked approximately a century after Lowell's 1895 book, *Mars,* began making big and controversial waves.

There's no doubt that the tubes—largely translucent or transparent, but not exclusively—exist, and in large amounts. And there's no doubt whatsoever that the nature of the tubes require a definitive answer as to what they are. Theories include (a) the highly ironic idea that Lowell and Schiaparelli were, mostly, correct from the very beginning—that the tubes were used by Martians to divert precious and vital water supplies to areas of the planet that were running perilously low when Mars began to degrade, whether due to catastrophic environmental developments or as a result of a planet-wide war, and (b) that the tubes were a means of transportation for the Martians themselves.

Jeffrey McCann, of the Research Abyss, said, somewhat echoing the theories of Percival Lowell: "A lot of people have seen these strange structures and have tried to speculate as to what they might be. *Some would say these are huge water ducts funneling water from one area to another* [author's emphasis], others are firm in thinking that these are some sort of natural geological anomaly only occurring on Mars."[2] As these words demonstrate, sometimes it's hard to fully move on from old theories.

Few people—if any—deny the undeniably odd appearance of the tubes: They look just like gigantic, ribbed earthworms. In fact, after what became famously known as the "glass tube photo" taken by the *Mars Global Surveyor* surfaced, Mars researcher Richard Hoagland highlighted the controversy on his website. It can be accurately said that Hoagland was *the* primary figure who brought the matter of the tubes to an eager and amazed audience. Ron Nicks, a geologist, recalled of one particular image sent back to Earth:

> This remarkable "tube," roughly a mile in length and hundreds of feet wide, appears to cling to a desert canyon wall near the canyon's bottom, and extend along its entire length. The feature has the appearance of being "translucent," of being supported at somewhat regular intervals by "ribs," and of being quite cylindrical—with a localized, internal structure at one point of considerably higher albedo (brightness). To define this feature in purely geological terms has been a considerable challenge.[3]

These ribs, or arches—the latter may be better and more correct terminology—play a significant role in the theories and conclusions of those who doubt that the tubes are merely natural phenomena. On this issue Richard Hoagland and Mike Bara, the authors of *Dark Mission: The Secret History of NASA,* said: "Critics in the past attempted to pass off similar arches as 'sand dunes.' To be sure, there are some superficial resemblances between these 'arches' (and similar structures near the base of some pyramids at Cydonia) and real sand dunes—but on any sort of close examination, the 'sand dunes' argument quickly falls apart."[4]

As for what the photo actually shows us, Glen Hiemstra, aka "The Futurist," echoed:

> The land area seen in the full image is about 20 miles by 2 miles. The portion of the tube exposed, more clearly seen in the blown-up

image, is about 1.5 miles long and perhaps 100 meters in diameter. It appears to be translucent, as though made of glass. It has circular markings, perhaps support structures, at regular intervals. The idea that it may be naturally occurring ice is apparently ruled out by the temperature at this region of Mars. So what is it?[5]

That was the question that just about every devotee of Martian puzzles wanted answered.

Stephen Wagner, writing in a 2019 article titled "The Most Mysterious Anomalies on Mars," noted: "McCann and Joseph P. Skipper, both researchers into Martian anomalies, call this image 'the real smoking gun as to life on Mars.' The image was discovered in June 2000 among the many images posted at Malin Space Science Systems, which has tens of thousands of Mars pictures available for online viewing and examination."[6]

Mac Tonnies, who believed that at least some of the anomalies on the surface of Mars were evidence that Mars was once inhabited by intelligent entities and that it was Earth-like in nature, was very much in two minds on the matter of the tubes' nature and purpose—if there even *was* a purpose to it all. In other words, Mac wasn't sure what the tubes were, or what they were used for. He told me that he was more inclined to think that the tubes had "geological" origins, rather than having been constructed by Martians, but he remained open-minded on the debate. Interestingly, Tonnies said that he received a degree of flak when he admitted that he wasn't totally on board with the theory that the tubes were the work of Martians. "I was spoiling their fun, is how I interpreted it," said Tonnies. He added: "The 'I want to believe' factor of *The X-Files* definitely took a big role in those early days when the tube pictures began [to appear]. The feedback I received when my book [*After the Martian Apocalypse*] was published made that obvious. I wasn't saying what they wanted to hear. Or, not enough of it."[7]

As time progressed, and as further revelations surfaced, Tonnies eventually made it abundantly clear that he viewed the tubes as genuine anomalies, but, when pushed, he backed away from fully endorsing the idea that what we're seeing are massive constructs—the remains of a long-gone civilization of extraterrestrial nature.

Today, when it comes to the matter, the makeup, and the origins of the "glass tubes of Mars," there is an undeniable division between the believers and the non-believers. The effects of nothing stranger than Mars's sand dunes? The incredible work of a race of Martians who fought valiantly to find ways to secure much-needed supplies of water, before time finally ran out for them? Whatever the answer to those questions, the fact is that the tubes remain an admittedly weird enigma—even to many of those who don't buy into aliens.

CHAPTER 5

"Is Mars Sending Us Signals?"

|||||||||||||||||||||||||||||||||||||||||

When it comes to the matter of searching for the Martians, sometimes we don't have to go looking for the evidence at all. On occasion, the Martians make it really easy: They contact us. At least, that's what we are led to believe has happened. Certainly, the most intriguing example of this situation revolved around a brilliant, maverick scientist named Nikola Tesla. Before we get to the matter of Martian–Tesla interaction, though, let us first take a look at the life and career of Tesla, who was certainly a man way ahead of his time. Tesla entered this world on July 10, 1856, in Smiljan, Lika (present-day Croatia). Notably, Tesla developed his passion for science and technology from his mother, Djuka; she had a flair for creating new and novel household appliances. Tesla's father, Milutin, had a career in the church: He was a Serbian Orthodox priest. Even at an early age, Tesla was demonstrating his fascination for science; as a teenager he entered the University of Prague and the Graz-based Polytechnic Institute. It quickly became apparent that Tesla's primary interest was the field of electricity. In his mid-20s, Tesla took a position with a Budapest-based phone company.

Things really changed for Tesla, however, when he moved to the United States; it was a giant leap that ensured great things for the man himself.

Tesla was responsible for some of the major developments in recent times. He continued to push the barriers of science, doing so pretty much right up to the time of his death in 1943—when many of his files were confiscated by the FBI. It should be stressed, though, that the FBI has now placed many of its files on Tesla and his work in the public

Nikola Tesla: in communication with Martians? (Wikimedia Commons)

domain. The provisions of the Freedom of Information Act made this happen. What, however, does all of this have to do with Mars—and, more specifically, the Martians? Let's take a look.

On January 4, 1901, the *San Francisco Examiner* newspaper ran an article with an incredible, lengthy headline. It read as follows: "World Speaks to World with Mysterious Signals through Vast Space—Tesla, The Electrician, Says He Received a Message from Mars." The article began: "Nikola Tesla has had the first call of the century from a neighboring planet. He has communicated with Mars, he declares while on Pike's Peak [Colorado], delving into the mysteries of the wireless transmission of electrical energy. The summons was faint, but, according to Tesla, not to be mistaken."[1]

In dramatic, lofty fashion, the *Examiner* left its audience something to muse upon: "A new voice from a planet, millions of miles removed, was spoken over one of the myriad of unwired telephones of the universe, and there, near the lonely mountain peak, in the fathomless calm of night, the voice at last found a listener and world spoke to world in language strange at first, but sure to be clearer, says Tesla, ere the Twentieth Century has finished its course."[2]

It's no surprise that the story provoked a great deal of amazement. Other newspapers picked up on the story—and in double-quick time, no less. Headlines practically screamed at their audiences: "The Distinguished Feeble Electrical Disturbances Which Could Not Have Been of Solar or Earthly Origin" and "The Discoverer Believes One of the Planets May Have Already Perfected a Scheme of Interplanetary Communication" were just two of many.[3]

It was the words of Tesla that really stood out, however:

As I have already said, one of the planets in the solar system may be ahead of us in the evolution. Their means of interplanetary communication may be perfect, but we have yet to learn their sign language. It is impossible at present, even to suggest a code and my observations to the Red Cross Society on New Year's Eve were purely speculative, but for purposes of illustration they will answer the purpose at present. It is enough to say at this time that a message from Mars which might be a triangle to them would appear as some other form to us, and vice-versa. These differences can only be reconciled by time and careful study. It is wonderful enough, is it not, that a beginning has been made.

On Pike's Peak I set out to carry on my experiments along three different lines: First, to ascertain the best conditions for transmitting power without wires; second, to develop apparatus for the transmission of messages across the Atlantic and Pacific oceans, and, third, to work on another problem, which involves a still greater mastery of electrical force. I consider of still greater importance than even the transmission of power without wires, and which I shall make known in due course.

In my laboratory in New York I was able to go only to electrical discharges of sixteen feet in length, and I had only reached effective electrical pressures of about 8,000,000 volts. To carry the problems further I had to master electrical pressures of at least

50,000,000 volts, and electrical discharges were necessary for some purposes measuring at least fifty or one hundred feet.

The results I attained were far beyond any I had expected to reach. I found that my mental vision was incomparably clearer, so much so that I could look back in thought to my laboratory in New York, and in examining familiar objects in the rooms there I could notice the smallest scratch on them, and in scanning the features of my assistants I could notice the slightest marks on their faces, as though they had been actually before me. Now in the city the mental images are much duller.

One of the first observations I made in Colorado was of great scientific importance and confirmatory of a result I had already obtained in New York. I refer to my discovery of the stationary electrical waves in the earth. The significance of this phenomenon has not yet been grasped by technical men, but it virtually amounts to a positive proof that, with proper apparatus, such as I have per-fected, a wireless transmission of signals to any point on the globe is practicable. When I read statements to the effect that such a thing is impossible and recall the numerous adverse criticisms of my expressed confidence that I can ultimately accomplish this, I experience a feeling of satisfaction.

As I think over it now it seems to me that only a man absolutely stricken with blindness, insensible to the greatness of nature, can hold that this planet is the only one inhabited by intelligent beings.

Was H. G. Wells's 1897 novel, *The War of the Worlds*, about to become a startling—and maybe even fear-filled—reality? Not every-one thought so. Among those was Professor A. S. Skinner, who was employed at the United States Naval Observatory. He said, succinctly: "This is the first I have heard of Tesla's latest accomplishment. You can rest assured it is imaginative and visionary. I do not care to make any further comment on it or the possibilities of the case."

Skinner, clearly, had very little time for Tesla.

Professor T. J. J. See, who also worked at the Observatory, said: "I have read the story about this discovery. I do not care to say anything about it." Clearly, Professor See, like A. S. Skinner, was no fan of Tesla.

See had far more to say:

He indicates that Mars might be the world which is attempting to communicate with us. It is most peculiar that any discovery involving Mars should be made just now. Once in every fifteen years that planet reaches its greatest distance from the earth, when it is 62,000,000 miles away. The nearest that it ever gets is 36,000,000 miles. It happens that just now Mars is at the farthest distance that it ever gets from us. Certainly it is hard to understand why, at such an inopportune time, signals from the Martians should have been detected by Tesla.

The professor concluded: "Mars may be populated. We cannot make any positive assertions regarding that, but it will take more than a mere statement from Nikola Tesla to prove the existence of signals from the planet to us."

Or, rather, "take that!"

There were also the words of yet another respected figure at the Observatory, Professor S. I. Brown: "Any proclamation coming from Tesla is likely to be received with considerable incredulity. I have been looking in vain for some practicable results to follow the extravagant claims that he has constantly made during the past half-dozen years. Most of his discoveries have materialized only in statements such as this which he gives out every once in a while for publication."

Not everyone was quite so quick to dismiss such matters, however. In May 1902, the 1st Baron Kelvin, born William Thompson, raised eyebrows aplenty when he gave his support to Tesla's claims and conclusions. The baron was certainly no fool: He was a well-respected, renowned engineer and physicist. A case of "gotcha!"? Maybe so.

On September 2, 1921, the *New York Tribune* ran an article that said, in part, none other than Guglielmo Marconi, a renowned Italian electric engineer, "believed he had intercepted messages from Mars during the recent atmospheric experiments with wireless on board his yacht *Electra* in the Mediterranean."[4] Incredibly, the media reported that Marconi came to suspect that the Martians were sending messages by nothing less than Morse Code. That wasn't entirely accurate, however; in part, it was down to tried and tested press sensationalism.

In her article "Everybody Shut Up! We're Listening to Mars," Jessica Leigh Hester puts matters right:

> In the early 1900s, Marconi began telling newspapers about "strange sounds" that he found in his transmissions. He imagined these to be "distinct, unintelligible" messages rather than wayward noise—they bore some similarity to the sound of the Morse-code "S" (dot-dot-dot)—and he attributed them to "the space beyond our planet." Newspapers quoted Marconi beside illustrations of pot-bellied, antenna-sporting Martians fiddling with the dials of their own radios beneath a canopy of stars and planets.[5]

On January 29, 1920, the *New York Times* published a feature on the Mars–Marconi controversy. In a wire from London, England, the *Times* said: "William Marconi informs *The Daily Mail* that investigations are in progress regarding the origin of mysterious signals which he recently described as being received on his wireless instruments. He hopes to make a statement on the subject at an early date. Marconi insists that 'nobody can yet say definitely whether they originate on the earth or in other worlds.'"[6]

The *New York Times* added:

> Is Mars sending us signals? That is the question for which a long time has interested us, since the publication of the Martian geo-

graphical charts, on which were observed singular features, the origin of which did not appear to be due merely to chance. We should be glad to take a step further toward our neighbors of the skies, who, perhaps for centuries have addressed to us signals to which we have never known how to reply, terrestrial humanity being still absorbed with the grosser demands of material affairs.[7]

Over time, Marconi backtracked on some of his thoughts concerning messages from Mars, but he never really dismissed the possibility that communications from our nearest planetary neighbor might have been a reality. Maybe they were. And, in the same way that we are seeking out the Martians, maybe they are doing likewise when it comes to us.

Now, let's take a look at the issue of Martians in the post–Second World War era.

CHAPTER 6

"Objects Sighted
May Possibly Be
Ships . . . from Mars"

||||||||||||||||||||||||||||||||||||||

It was on June 24, 1947, that the UFO/Flying Saucer phenomenon began. That was the date on which an American pilot named Kenneth Arnold encountered a squadron of strange-looking aircraft flying near Mount Rainier, Washington State. In Arnold's own words:

> On June 24th, Tuesday, 1947, I had finished my work for the Central Air Service at Chehalis, Washington, and at about two o'clock I took off from Chehalis, Washington, airport with the intention of going to Yakima, Wash. My trip was delayed for an hour to search for a large marine transport that supposedly went down near or around the southwest side of Mt. Rainier in the state of Washington and to date has never been found.[1]

Arnold continued:

There was a DC-4 to the left and to the rear of me approximately fifteen miles distance, and I should judge, at 14,000 foot elevation. The sky and air was clear as crystal. I hadn't flown more than two or three minutes on my course when a bright flash reflected on my airplane. It startled me as I thought I was too close to some other aircraft. I looked every place in the sky and couldn't find where the reflection had come from until I looked to the left and the north of Mt. Rainier where I observed a chain of nine peculiar looking aircraft flying from north to south at approximately 9,500 foot elevation and going, seemingly, in a definite direction of about 170 degrees.

A puzzled Arnold recalled:

I thought it was very peculiar that I couldn't find their tails but assumed they were some type of jet plane. I was determined to clock their speed, as I had two definite points I could clock them by; the air was so clear that it was very easy to see objects and determine their approximate shape and size at almost fifty miles that day. I remember distinctly that my sweep second hand on my eight day clock, which is located on my instrument panel, read one minute to 3 P.M. as the first object of this formation passed the southern edge of Mt. Rainier.

We'll end with these words from Arnold:

As the last unit of this formation passed the southern-most high snow-covered crest of Mt. Adams, I looked at my sweep second hand and it showed that they had travelled the distance in one minute and forty-two seconds. Even at the time this timing did not upset me as I felt confident after I would land there would be some explanation of what I saw.

History has shown that Arnold's experience of June 24, 1947, never was resolved to the satisfaction of everyone: the military, the media, and the public. All we can say for sure is that Arnold's encounter provoked a huge wave of UFO activity in the summer of 1947—and it showed no signs of stopping. The era of the UFO was born. And it still lives.

Although the UFO controversy began in the summer of 1947, it's a fact that encounters with alleged aliens in that era were all but nonexistent. In fact, it wasn't until the early 1950s that people began to make claims to the effect that they had undergone face-to-face encounters with aliens. In nearly all cases, the aliens were very human-like. The only differences were that they sported heads of long blond hair, which, of course, was hardly the style for men in early-1950s-era USA. In that sense, they really stood out. But, trim their hair and they would look just like us. The extraterrestrials soon became known in the field of Ufology as the Space Brothers, while those that encountered the beings from beyond were dubbed the Contactees. Unlike today's aliens—bug-like, dwarfish, black-eyed things who routinely abduct people in the dead of night and in trauma-filled fashion—the Space Brothers were friendly beings whose main role seemed to be to warn people of the perils of nuclear weapons. Not only that: Many of the Contactees claimed that their brothers and sisters from the heavens above were Martians. As a consequence, beings from the Red Planet had been hauled into the growing UFO controversy.

The Contactees claimed that they met the Space Brothers at isolated, lonely locales, such as the deserts of California, New Mexico, and Arizona. Not only that, the Space Brothers urged those they targeted for recruitment to go out and spread the words and advice of the aliens. They certainly did that. In no time at all, the Space Brother/Contactee phenomenon became the dominating aspect of early 1950s Ufology. Without doubt the most famous (many would soon say infamous—and many still do) of all the Contactees was

George Adamski. His claims of encounters with benign, human-like aliens captured the attention and imagination of the public to a major degree. For example, his first book, *Flying Saucers Have Landed*, cowritten with Desmond Leslie, was a huge seller: Sales reached no less than six figures.[2]

Following in the footsteps of Adamski were Truman Bethurum, who in 1954 wrote *Aboard a Flying Saucer*, an entertaining saga of Bethurum's alleged encounters with a hot space-babe known as Captain Aura Rhanes; George Hunt Williamson, who was very much in the style of Adamski; and Orfeo Angelucci, who was granted flights on alien saucers and became a fixture on the UFO-based lecture circuit. While there were certain differences between the various tales (or the yarns) that the Space Brothers told, there was one theme that really stood out: the claim that many of the aliens came from Mars, or had connections to Martians. Regardless of whether or not one buys into the often very tall tales of the Contactees, it was the incredible influence of these admittedly gifted storytellers that led many to look toward Mars for answers concerning the UFO presence on our world.

Even certain elements of the US military found itself "infected" by such stories. In the summer of 1952, Commander Randall Boyd, of US Air Force Intelligence, quietly advised N. W. Philcox, at the time the FBI's liaison with the Air Force, that "It is not entirely impossible that the objects sighted may possibly be ships from another planet such as Mars."[3]

All of this brings us to a man named George Van Tassel. As a result of his claimed encounters with aliens in the early part of the 1950s, Van Tassel developed a distinct aversion to nuclear weapons and became the subject of an FBI surveillance file that ran to almost four hundred pages. It was while Van Tassel and his family were living near Landers, California, in the Fifties (in a hollowed-out cave, below a giant rock known locally and appropriately as "Giant Rock"; and, no I'm not making this up) that Van Tassel's encounters with those long-haired extraterrestrials began. The

The Martians in the mid-twentieth century. (Nick Redfern)

Space Brothers would often meet with Van Tassel late at night and out in the dark desert. Love and peace, nuclear disarmament, and even the secrets of immortality were discussed.[4]

Whereas Adamski, Bethurum, and the rest were content to write books, Van Tassel did something very different: At the urging of the Space Brothers, he put on a yearly, outdoor UFO event at Giant Rock. So successful were the events that at their height Van Tassel's gigs had audiences in the figure of about *eleven thousand,* which massively eclipses anything that can be seen on the UFO lecture circuit today.[5] It's no wonder, then, that all of this talk of getting rid of our nukes and living in harmony with the Russians attracted the attention of J. Edgar Hoover's agents.

George Van Tassel's FBI file makes for fascinating reading, partly because the Bureau was deeply interested in certain UFO phenomena—of the "Ancient Aliens" type, no less—that Van Tassel often lectured on. And also on topics that had a bearing on the story that this book tells, as we shall now see. In April 1960, a report was prepared by a special agent of the FBI who had clandestinely sat in the audience of one of Van Tassel's lectures in Colorado. The following is the complete text of the relevant document[6]:

On April 17, 1960, a lecture was given at Phipps Auditorium, City Park, Denver, Colorado, which was advertised to be a lecture, movie film, and discussion of unidentified flying objects. The audience was comprised of a majority of older individuals and also a majority of the audience was female. There were few young people, although some family groups.

The program was sponsored by the Denver [deleted] one of which meets monthly at the [deleted], Lakewood, Colorado, whose [deleted] was the Master of Ceremonies. The program consisted of a 45 minute movie which included several shots of things purported to be flying saucers, and then a number of interviews with people from all walks of life regarding sightings they had made of such unidentified flying objects. After the movie [deleted] gave a lecture which was more of a religious-economics lecture rather than one of unidentified flying objects.

[Deleted] stated that he had been in the "flying game" for over 30 years and currently operates a private Civil Aeronautics Authority approved airfield in California. He said he has personally observed a good many sightings and has talked to hundreds of people who have also seen flying saucers. He said that he has also been visited by the people from outer space and has taken up the cause of bringing the facts of these people to the American people. He said it is a crusade which he has undertaken because he is more or less retired, his family is grown and gone from home, and he feels he might be doing some good by this work.

The major part of his lecture was devoted to explaining the occurrences in the Bible as they related to the space people. He said that the only mention of God in the Bible is in the beginning when the universe was being made. He said that after that all references are to "out of the sky" or "out of heaven." He said that this is due to the fact that man, space people, was made by God and that in the beginning of the world the space people came to the earth and left animals here. These were the pre-historic animals which existed at a body temperature of 105 degrees; however, a polar tilt occurred whereby the poles-shifted and the tropical climates became covered with ice and vice versa.

He said that then the space people again put animals on the earth and this is depicted in the Bible as Noah's ark. He said that after the polar tilt the temperature to sustain life was 98.6 degrees, which was suitable for space people, so they established a colony and left only males here, intending to bring females at a later date on supply ships. This is reflected in ADAM's not having a wife. He said that ADAM was not an individual but a race of men. He said that this race then inter-married with "intelligent, upright walking animals," which race was EVE.

Then when the space people came back in the supply ships they saw what had happened and did not land but ever since due to the origin of ADAM, they have watched over the people on earth. He said that this is recorded in the Bible many times, such as MOSES receiving the Ten Commandments. He said the Ten Command-ments are the laws of the space people and men on earth only give them lip service. Also, the manna from heaven was bread supplied by the space people.

He also stated that this can be seen from the native stories such as the Indians in America saying that corn and potatoes, unknown in Europe, were brought here by a "flaming canoe." He said this refers to a space ship and the Indians' highest form of transportation was the canoe, so they likened it unto that. He said this can be shown also by the old stories of Winged Chariots and Winged White Horses, which came from out of the sky. He said that JESUS was born of MARY, who was a space person sent here already pregnant in order to show the earth people the proper way to live. He said the space people have watched over us through the years and have tried to help us. He said they have sent their agents to the earth and they appear just as we do; however, they have the power to know your thoughts just as JESUS did. He said this is their means of communication and many of the space peo-ple are mute, but they train a certain number of them to speak earth languages.

He said that the space people here on earth are equipped with a "crystal battery" which generates a magnetic field about them which bends light waves so that they, the space people, appear invisible. He said this has resulted in ghost stories such as footsteps, doors opening, and other such phenomena.

He stated that the space people are now gravely concerned with our atom bombs. He said that the explosions of these bombs have upset the earth's rotation and, as in the instance of the French bomb explosion in North Africa, have actually caused earthquakes. He said that the officials on earth are aware of this and this was the reason for the recent Geophysical Year in order to try to determine just what can be done. He said these explosions are forcing the earth toward another polar tilt, which will endanger all mankind. He said that the space people are prepared to evacuate those earth people who have abided by the 'Golden Rule' when the polar tilt occurs, but will leave the rest to perish.

He advised that the space people have contacted the officials on earth and have advised them of their concern but this has not been made public. He also said that the radioactive fallout has become extremely dangerous and officials are worried but each power is so greedy of their own power they will not agree to make peace.

[Deleted] also spent some time saying that the U. S. Air Force, who are responsible for investigations of unidentified flying objects, has suppressed information; and as they are responsible only to the Administration, not to the public, as elected officials are, they can get away with this. He said that also the Air Force is afraid that they will be outmoded and disbanded if such information gets out.

He said that the Administration's main concern in not making public any information is that the economy will be ruined, not because of any tear that would be engendered in the public. He said this is due to the number of scientific discoveries already made and that will be made which are labor saving and of almost permanency so that replacements would not be needed.

In summation, [deleted] speech was on these subjects: (1) Space people related to occurrences in Bible. (2) Atom bomb detrimental to earth and universe. (3) Economy is poor and would collapse under ideas brought by space people. Throughout his lecture, he mentioned only the U.S. economy and Government and the US Air Force. He did refer to the human race numerous times but all references to Government and economy could only be taken as meaning the U. S. One question put to him was whether sightings had been made in Russia or China. He answered this by saying sightings had been reported all over the world, but then specifically mentioned only the U. S., Australia, New Zealand, and New Guinea.

He also mentioned that he was not advocating or asking for any action on the part or the audience because he said evil has a way of destroying itself. He did say that he felt that the audience, of about 250 persons, were the only intelligent people in Denver and he knew they had not come out of curiosity but because they wanted to do the right thing. He said that they were above the average in intelligence and when the critical time came, the world would need people such as this to think and guide.

It goes without saying that much of what Van Tassel was stating in 1960—and that he had been saying as far back as 1952—is just the kind of material that one can regularly see on the History channel's show *Ancient Aliens:* worldwide disaster in the past, alien intervention, the genetic altering of early humans by sexual intercourse, and more. If nothing else, Van Tassel was a definitive trendsetter. He was someone whose views on ancient extraterrestrials visiting our world thousands of years ago came long before the theories of Erich von Däniken, whose *Chariots of the Gods?* became a huge bestseller after its publication in 1968. (Yes, the title of von Däniken's book does have a question mark after it, even though it's so often omitted.)

Not only that: The very year in which the FBI took careful notice of Van Tassel's accounts of aliens meddling with the human race thousands of years ago—1960—was also the year the Brookings Institution prepared a report for NASA titled "Proposed Studies on the Implications of Peaceful Space Activities for Human Affairs." In part, the report states: "While face-to-face meetings with [alien life] will not occur within the next twenty years (unless its technology is more advanced than ours, qualifying it to visit earth), artifacts left at some point in time by these life forms might possibly be discovered through our space activities on the Moon, Mars, or Venus."[7]

That, way back in 1960, George Van Tassel was loudly, and widely, sharing his thoughts and conclusions on some of the very things that appear later in this book, and that the Brookings Institution—in the very same year—was discussing certain "artifacts" that might have a link to Mars, is both intriguing and extraordinary.

CHAPTER 7

"A Visual History
of a Race's
Heroic Death"

||||||||||||||||||||||||||||||||||

In one of those "you can't make this stuff up" situations, it's time to turn our attentions to the connection between the Face on Mars and a legendary comic book artist, a man who turned out to be one of the key figures in the "superhero" genre. Yes, really. Who was that man? None other than Jack Kirby, who was born in 1917 and who became famous for his artwork for Marvel Comics, DC Comics, and Harvey Comics. Kirby, in 1940, was the cocreator with Joe Simon of Captain America, who first appeared in Timely Comics. Characters that Kirby created with Marvel's Stan Lee in the early 1960s included the Fantastic Four, the Hulk, the X-Men, and Iron Man. They have all appeared in megabucks movies, with the *X-Men* and *Iron Man* movies being particularly successful. Kirby chose to move to DC Comics in 1970, majorly angered due to his belief that his specific place in the creation of these—and more—characters had been deliberately played down by Marvel Comics. Kirby died in 1994, at the age of 76. But, it's to 1958 that we now have to turn our attentions.

It was in that year that Harvey Comics published a three-part production that went by the title of *Race for the Moon*. Although Kirby is primarily known for his distinct, and easily recognizable, style of drawing, he wasn't just the artist on two issues of *Race for the Moon*. He was the writer of the story, too. Of the three stories, it's the second comic book in the trilogy that, for us, is the most important. Just why might that be? Because it revolves around nothing less than a giant, carved, stone head that is found on the surface of the planet Mars. It's even referred to as "the Face on Mars" in the comic book. A strange, but wholly coincidental precursor for what was to come to the fore a couple of decades later? Maybe so. On the other hand, though, perhaps there is far more to all of this than meets the eye—something that revolves around nothing less than government secrecy and conspiracy, as we shall soon see. Let's begin with the plot of Kirby's tale.

The story is an entertaining one and revolves around a group of astronauts who embark on an ambitious mission to Mars. Only when they reach the Red Planet, to their amazement and excitement, does the group come to realize that, while Mars is clearly dead, it was most assuredly not always that way. The adventuring astronauts are confronted by a gigantic stone creation, which is clearly not the work of nature. Yes, you've guessed it right: They stumble on a massive, carved head of human-like proportions. It dominates the ravaged, Martian landscape. The hero of the story, astronaut Ben Fisher, decides to climb the massive creation. In doing so, he discovers, to his excitement, that the carved-out eyes of the construction are really something else entirely: They're nothing less than openings to an ancient world that is hidden from view—that is, until Fisher carefully descends into those stone eye-sockets and discovers the shocking truth.

In quick time, Fisher passes out. In his unconscious state, Fisher has a graphic vision of Mars in times long gone: of a thriving world and of an advanced culture of giant humanoids (in later chapters we'll see how tales of Martian giants surface regularly)—a civilization, perhaps, not too dissimilar to our own. That is, until something terrible happens: something that Fisher gets to "see" for himself. Mars is suddenly attacked by the powerful, ruthless denizens of an ancient world that exists between Jupiter and Mars. They turn out to be the archenemies of the Martians. In no time at all, Mars is laid to waste; civilization is decimated, culture is utterly shattered, and a world is soon dead. Things aren't quite over, though: A surviving Martian is able to utterly destroy the world of the creatures that attacked them. That same world is then blasted into pieces, becoming what we, today, call the Asteroid Belt. Rather notably, we'll see later how a noted psychic, in 2019, made a connection between the demise of the Martian race and the Asteroid Belt—something that adds yet another layer of intrigue to the overall story of Mars.

With the war over—one world now bereft of life and the other completely destroyed—Fisher begins to wake up from his unconscious state and quickly realizes that what he saw on passing out was actually a brief, almost magical, glimpse into the distant past. It's as if somehow, as Fisher states, he got to see "a visual history of a race's heroic death" and experienced a strange, but amazingly accessible, "surviving memory."[1] And the story doesn't end there, as Christopher Knowles, of the Secret Sun blog, demonstrates: "Shortly before Kirby wrote 'The Face on Mars,' he wrote another, similar story called 'The Great Stone Face,' about an ancient astronaut cargo cult (way, way, way before such things were fashionable in comics). 'The Face on Mars' seems very much to be a sequel/companion to this story, as it covers many of the same bases."[2]

There's no doubt that there are undeniable parallels between the Face on Mars as we know it today and the creation of Jack Kirby's fictional Face on Mars, back in the late 1950s. Coincidence? Such a thing is certainly not impossible. On the other hand, there may well be a more plausible—but amazing and incredible—answer to the admittedly strange situation we find ourselves in. We find the answer to all of this in the world of one Tom Corbett.

Just like Ben Fisher in *Race for the Moon,* Tom Corbett was a fictional, hero-type creation. The primary figure in the creation and expansion of Tom Corbett was sci-fi legend Robert Heinlein. His books included *Space Cadet* (1948), *Stranger in a Strange Land* (1961), and *Podkayne of Mars* (1963). So popular was Heinlein's 1948 novel, there were loud calls for more adventures of Tom Corbett. The publisher, Grosset & Dunlap, was more than happy to oblige fans: Eight books duly followed, with the serial finally ending in 1956. There were other adventure-filled spinoffs, however: both a comic book and a 3D "View Master." The latter is particularly jaw-dropping; the reason being that—as with Jack Kirby's 1958 work—the View Master version of the Tom Corbett stories seemed to eerily preempt what is being discussed today in relation to Martian anomalies on the surface of the planet. In the three-part, Kodachrome picture-based story, we see our astronauts head out to Mars where they find a gigantic, ancient pyramid and a carved face. Yes, things are getting stranger by the minute. It's almost as if Jack Kirby and the team behind the *Space Cadet* phenomenon knew something that the rest of us wouldn't even begin to be exposed to until the 1970s. How could such a thing be? That's the $64,000 question that still has to be answered.

It's important to note that aside from Robert Heinlein, there were others who worked on the Tom Corbett franchise. They included Willy Ley and Wernher von Braun. The former was a German–American expert in the field of rocketry and science. His books included *Mars and Beyond, Mariner IV to Mars,* and *The Exploration of Mars.* Ley clearly had a deep interest in Mars. He also did classified work for the US government, also in the domain of rocket science.

As for von Braun, he too was someone who worked in the field of rockets. In fact, during the Second World War, von Braun—a through-and-through Nazi—perfected the V-2 rockets that wrought massive damage on the UK. Outrageously, given his background as a Nazi, von Braun was brought to the United States in 1945 under the US government's secret Operation Paperclip[3], which sought to grab as many German scientists as possible, after the war was over. Even more outrageously, von Braun was given the position of director of NASA's Marshall Space Flight Center. Like Ley, much of von Braun's work was shrouded in secrecy, particularly so in the late 1940s and 1950s. Interestingly, in 1948 von Braun wrote a book titled *The Mars Project*. Von Braun's writing makes it clear he was highly enthused by the idea of manned missions to Mars. It's no surprise, then, that both Ley and von Braun were consulted by the Tom Corbett/*Space Cadet* team. Things don't end there.

It was while the *Space Cadet* phenomenon was still riding high that von Braun was brought into the fold of none other than Walt Disney. There was a good reason for this: The Disney Corporation was planning on making a three-part production on the world of outer space. They were *Man in Space, Man and the Moon,* and *Mars and Beyond*. They were incredibly popular with the US population. None other than President Dwight D. Eisenhower was a huge fan of the three shows, believing that they were directly instrumental in getting the United States galvanized into aiming for the Moon and Mars.[4] It was while he was working with Walt Disney that von Braun met Ward Kimball.

A high-up figure in the Disney Corporation (he worked on *Snow White and the Seven Dwarfs, Pinocchio,* and many others), Kimball, in 1953, was secretly hired by the CIA to work on the agency's UFO investigative program, the Robertson Panel, named after a respected physicist, Howard Robertson, who ran the prestigious group of scientists and psychologists. The Robertson Panel brought Kimball on board for two reasons: (1) Kimball had a deep interest in, and knowledge of, the UFO phenomenon, and (2) the CIA wanted to use the Walt Disney Corporation to try and manipulate the public's views on UFOs, no less. And how would they do that? By making a series of documentaries on UFOs, that's how. The UFO shows, ultimately, were scrapped. The rumor was that the CIA was unsure how to try and direct the mindset of the American people when it came to UFOs. It's important to note, however, the incredible threads that exist here.

We have Wernher von Braun, who had a deep interest in Mars and worked with the Walt Disney Corporation on space-based TV shows. We have von Braun, who mingled with Ward Kimball and the CIA's UFO-themed Robertson Panel. We have Willy Ley, who, like von Braun, was also fascinated by Mars. We have images of a pyramid and a carved face on Mars appearing in the Tom Corbett stories. And we have Jack Kirby who, in a strange and roundabout way, *just happened to do secret work for the CIA*. Seriously, yes.

Back in 1979, there was a major head-to-head between the United States and the government of Iran: Staff of the American Embassy in Iran were held by Iranian forces for a considerable amount of time. Six of them, however—State Department staff— had managed to avoid capture and were able to escape the wrath of the Iranians. Much of it was all thanks to the CIA and Jack Kirby. The CIA had an idea to get the six out of Iran. It was a very good idea, one that, as history has demonstrated, worked spectacularly well. CIA personnel made a bunch of bogus passports and fitted the group with wigs and makeup. The plan was to get the group out of Iran by having them pose as a team of Canadian movie-makers. If they were stopped at Tehran's airport, the plan was to say they were working on a movie adaptation of Roger Zelazny's 1968 novel, *Lord of Light*. To boost this scenario, some of Jack Kirby's artwork—specifically that which was designed for the *Lord of Light* project—was used to make the Iranians believe that the group really was Canadian, was working on a movie, and was not American Embassy staff trying their best to escape the Iranian regime. The ruse worked: The group carefully and stealthily made their way through the airport and back to the United States.

It is all of these conspiratorial connections—from Jack Kirby and his comic book–based face on Mars to the CIA, from the secret work of Wernher von Braun to Willy Ley, and from the Tom Corbett stories to tales of faces and pyramids on Mars—that have led to the theory that someone knew a great deal more about Mars and its anomalies long before the rest of us did. Think it couldn't be so? It actually could. Read on:

The National Investigations Committee on Aerial Phenomena— better known as NICAP—was one of the most influential of all the UFO research groups of all such groups that existed from the 1950s to the 1970s. In December 1960, NICAP ran an eye-catching article in its newsletter that was headed as follows: "Space-Life Report Could be Shock." NICAP expanded on this[5]:

The discovery of intelligent space beings could have a severe effect on the public, according to a research report released by the National Aeronautics and Space Administration. The report warned that America should prepare to meet the psychological impact of such a revelation. The 190-page report was the result of a $96,000 one-year study conducted by the Brookings Institution for NASA's long-range study committee.

NICAP continued:

Public realization that intelligent beings live on other planets could bring about profound changes, or even the collapse of our civilization, the research report stated. "Societies sure of their own place have disintegrated when confronted by a superior society," said the NASA report. "Others have survived even though changed. Clearly, the better we can come to understand the factors involved in responding to such crises the better prepared we may be." Although the research group did not expect any immediate contact with other planet beings, it said that the discovery of intelligent space races "could nevertheless happen at any time."

As for the title of the report that had got NICAP in such a whirl, it was "Proposed Studies on the Implications of Peaceful Space

Activities for Human Affairs." It was the product of the Brookings Institution. The group's website tells us the following: "The Brookings Institution is a nonprofit public policy organization based in Washington, DC. Our mission is to conduct in-depth research that leads to new ideas for solving problems facing a society at the local, national and global level."[6] It had been commissioned by NASA's Committee on Long Range Studies.

The report covers numerous aspects of "space activities," but one section—albeit a small section—really stands out. It states the following attention-grabbing words with respect to possible encounters with an extraterrestrial race: "While face-to-face meetings with it will not occur within the next twenty years (unless its technology is more advanced than ours, qualifying it to visit earth), *artifacts left at some point in time by these life forms* [author's emphasis] might possibly be discovered through our space activities on the Moon, Mars, or Venus."[7]

That reference to "artifacts" is most intriguing. Was this particular sentence deliberately inserted into the report, in 1960, to get us ready for the day when theoretical alien artifacts suddenly become all too real extraterrestrial artifacts—such as carved heads and massive pyramids, and all to be found on the surface of Mars? Just maybe Brookings and NASA already knew something of what was on Mars, even in the 1950s. If so, that might very well explain the undeniable high strangeness that dominates this chapter and why, perhaps, some arm of government wanted the public—in the Fifties—to slowly become acclimatized to imagery of pyramids, huge carved heads, and an ancient Martian civilization. It may have been a plan that also involved the surfacing of the photos of the Face on Mars in the late 1970s.

Now, let's leave the latter part of the Seventies—a decade in which public Face on Mars research was in its infancy—and take a leap forward to 1984 and the CIA's connections to the Red Planet and a doomed race of Martians in a time frame long gone.

CHAPTER 8

"They're Ancient People. . . . They're Dying."

|||||||||||||||||||||||||||||||||||||||

May 22, 1984, was one of the most significant of all days in the incredible quest to seek out the truth of intelligent, alien life on Mars. So, why weren't we, the public, told all about it? How come the story wasn't splashed across the front pages of the world's newspapers? Why weren't major TV news networks on top of it? The answer to all of those questions is both controversial and amazing: The entire affair was deliberately shrouded in blankets upon blankets of secrecy. And who was responsible for ensuring that the secrecy remained intact? None other than the Central Intelligence Agency—the CIA. Operating out of its headquarters at Langley, Virginia, the CIA—for reasons that, decades later, still remain unclear—decided that it was vital to try to determine if Mars was a dead world, a planet teeming with life, or somewhere in between. It has to be said that the CIA took a most controversial and alternative way to securing the answers. The CIA's agents didn't employ the help of NASA or of any other space agency. In fact, rather astonishingly, and taking into consideration the subject matter (life on Mars), there is no evidence that NASA had *any* idea, whatsoever, of what the CIA was doing behind the curtains. Rather, the CIA used a process known as *remote viewing* or, in simple terms, spying of the psychic kind.

Approved For Release 2000/08/08 : CIA-RDP96-00788R001900760001-9

SUB: ......I'm seeing ah....It's like a perception of a shadow of people, very tall...thin, it's only a shadow. It's as if they were there and they're not, not there anymore.

MON;   Go back to a period of time where they are there.

SUB: ....Um.....(mumble) It's like I get a lot of static on a line and everything, it's breaking up all the time, very fragmentary pieces.

MON: Just report the raw data, don't try to put things together, just report the raw data.

SUB: I just keep seeing very large people. They appear thin and tall, but they're very large. Ah...wearing some kind of strange clothes.

MON: All right, now holding in this time period, holding in this time period, I want to move from your physical location in space to another physical location, but in this time period. Move now to:

      46.45 north
      353.22 east

Move in this time to:

      46.45 north
      353.22 east

SUB: ....Deep inside of a cavern, not a cavern, more like canyon. Um, I'm looking up, up the sides of a steep wall that seem to go on forever. And there's like ah... a structure with a...it's like the wall of the canyon itself has been carved. Again I'm getting a very large structures, no.... ah....no intricacies, huge sections of smooth stone.

MON: Do the structures have insides and outsides?

SUB: ....Yes, they're very, it's like a rabbit warren, corners of rooms, they're really huge, I don't, feel like I'm standing in one it's just really huge. Perception is that the ceiling is very high, walls very wide.

MON: (Real time plus 22 minutes.)* Yes that would be correct. All right, I'd like to move now to another location nearby. All right, move from this point in this time to:

Approved For Release 2000/08/08 : CIA-RDP96-00788R001900760001-9

The CIA secretly searches for the Martians. (CIA)

The International Remote Viewing Association describes this incredible skill as follows:

> Remote viewing is a mental faculty that allows a perceiver (a "viewer") to describe or give details about a target that is inaccessible to normal senses due to distance, time, or shielding. For example, a viewer might be asked to describe a location on the other side of the world, which he or she has never visited; or a viewer might describe an event that happened long ago; or describe an object sealed in a container or locked in a room; or perhaps even describe a person or an activity; all without being told anything about the target—not even its name or designation.[1]

It's hardly surprising that intelligence agencies, such as the CIA, would be highly enthused by the idea of using the psychic power of the human mind to spy on their enemies—maybe even on their allies, too. By the CIA's own admission, their interest in using psychic phenomena to uncover KGB agents in the United States, to locate Russia's hidden missile bases, and even to try to "read" the contents of some of the Kremlin's most secret files, goes back a very long time. We know this thanks to a man named Kenneth A. Kress.

In 1977, Kress, who, at the time, was an engineer in the CIA's Office of Technical Services (OSS), wrote a report on the potential use of supernatural phenomenon in the field of espionage. Its title: *Parapsychology in Intelligence.* In part, Kress stated:

> Anecdotal reports of extrasensory perception capabilities have reached U.S. national security agencies at least since World War II, when Hitler was said to rely on astrologers and seers. Suggestions for military applications of ESP continued to be received after World War II. In 1952, the Department of Defense was lectured on the possible usefulness of extrasensory perception in psychological warfare. In 1961, the CIA's Office of Technical Services became interested in the claims of ESP. Technical project officers soon

contacted Stephen I. Abrams, the Director of the Parapsychological Laboratory, Oxford University, England. Under the auspices of Project ULTRA, Abrams prepared a review article which claimed ESP was demonstrated but not understood or controllable.[2]

Kress was careful to note: "The report was read with interest but produced no further action for another decade."[3] That is true: Years went by before the CIA began to take a serious look at Remote Viewing and its potential. But, when they embraced it, they *really* embraced it.

In April 1972 a classified meeting, to discuss setting up a program, occurred between representatives of the OSS and laser physicist Dr. Russell Targ. The CIA liked what they heard from Targ: the feasibility of using psychics to help keep, primarily, the Soviet Union at bay and to dig out its most deeply hidden secrets via mind-power. There was, however, a problem: It revolved around Kenneth Kress's statement that Remote Viewing was not wholly "understood or controllable." While those attached to the program were in little doubt about its reality and importance, matters proved to be very much of a "hit and miss" variety. On some occasions the CIA's trained Remote Viewers had spectacular successes. On other occasions, though, the projects were blighted by failures. For the CIA, though, any kind of success was seen as a good one. This was understandable, given the fact that the Cold War was still very much a big, dangerous problem at the time. In the years ahead, the program was significantly expanded and, eventually, had dozens of Remote Viewers on its payroll. And amid all of the paranormal projects aimed at those nations seen as significant threats to the United States, a decision was quietly made by CIA personnel to do something radically different: *to go to Mars*. In a decidedly strange fashion, at least. It wasn't quite along the lines of "That's one small step for a man, one giant leap for mankind." It was, however, pretty damn close.

Before we go any further, an important question needs to be asked: How do we know, today, that Mars was a target of interest for the spies of the CIA? The answer is very simple. In 2000, the CIA quietly declassified into the public domain one of the 1984-era documents concerning the quest to find out if Martians really exist.[4] Or had lived hundreds of thousands of years ago or more. The document didn't create a huge firestorm of controversy at the time, but it certainly has in more recent times. However many other reports on the Remote Viewing of Mars may exist is anyone's guess. It should be stressed, though, that the single report we *do* have reveals a tumultuous and ultimately tragic story. It's a story that tells of an incredibly ancient race of Martians whose civilization is teetering on overwhelming extinction. Not only that, their home planet of Mars is spiraling into irreversible, global destruction. A once thriving world is about to be plunged into chaos and with no sign of turning the clocks back.

For the CIA, unchartered territory was about to become chartered. As the lyrics of "One Million Years B.C." by the Misfits come to mind, the CIA was heading off to seek out a Martian civilization.

For such a sensational subject, the CIA's file on Mars and its potential inhabitants has a wholly *un*sensational title: "MARS EXPLORATION MAY 22, 1984." And that's all. Two people take part in the mission to Mars. One is described as the subject and the other is referred to as the monitor. The document does not reveal their names; however, we have one of the two to thank for eventually coming forward and revealing his role in this extraordinary operation. That man was a skilled Remote Viewer: F. Holmes "Skip" Atwater. He was a captain in the US Army and the man who ran the Army's Remote Viewing (RV) program. Yes, multiple agencies of the US government decided to get into the world of Remote Viewing. It wasn't just the CIA, though. There was, however, a fair degree of cross-pollination that allowed agencies to share data and work with each other on top secret RV operations.

There is some debate as to who the other man was. It has been suggested that he was Joe McMoneagle, also an RV expert. McMoneagle, however, denied this, stating that the man was Robert Monroe, who founded the Monroe Institute. Monroe had a particular interest in the field of paranormal experiences and wrote such books as *Journeys out of the Body* and *Ultimate Journey*. With that said, let's now take an uninterrupted look at the transcript of the CIA document and the journey back in time to Mars, circa one million years BC, which, for a reason unknown, was the specific timeframe the CIA was interested in.[5] The viewer is referred to as the "SUB" (subject); the monitor of the viewing is detailed as the "MON" (monitor).

Method of site acquisition:
Sealed envelope coupled with geographic coordinates.

The sealed envelope was given to the subject immediately prior to the interview. The envelope was not opened until after the interview. In the envelope was a 3 X 5 card with the following information:

The planet Mars.

Time of interest approximately 1 million years B.C.

Selected geographic coordinates, provided by the parties requesting the information, were verbally given to the subject during the interview.

MON: (ROJ for 5/22 (May 22nd), time 10:09 AM.)

MON: (Plus 10 minutes, ready to start.) All right now, using the information in the envelope I've provided, exclusively focusing your attention now, using the information in the envelope, focus on: 40.89 degrees north 9.55 degrees west.

SUB: I want to say it looks like ah, I don't know, it sort of looks . . . I kind of got an oblique view of a, ah, pyramid or pyramid form. It's very high, it's kind of sitting in a large depressed area.

MON: Alright.

SUB: It's yellowish, ah, okra colored.

MON: All right. Move in time to the time indicated in the envelope I've provided you and describe what's happening.

SUB: I'm tracking severe, severe clouds, more like dust storm, ah, its geologic problem. Seems to be like a, ah. . . . Just a minute, I've got to iron this out. It's really weird.

MON: Just report your raw perceptions at this time, you're still early in the session.

SUB: I'm looking at, at a [sic] after effect of a major geologic problem.

MON: Okay, go back to the time before the geologic problem.

SUB: Um, total difference, it's ah . . . before there's no ah . . . ah I don't know . . . oh hell, it's like mountains of dirt appear and then disappear when you go before. See ah . . . large flat surfaces, very ah . . . smooth . . . angles, walls, they're really large though, I mean they're megalithic, ah . . .

MON: All right. At this period in time now before the geologic activity, look around, in and around this area and see if you can find any activity.

SUB: I'm seeing ah. . . . It's like a perception of a shadow of people, very tall . . . thin, it's only a shadow. It's as if they were there and they're not, not there anymore.

MON: Go back to a period of time where they are there.

SUB: Um . . . (mumble). It's like I get a lot of static on a line and everything, it's breaking up all the time, very fragmentary pieces.

MON: Just report the raw data, don't try to put things together, just report the raw data.

SUB: I just keep seeing very large people. They appear thin and tall, but they're very large. Ah . . . wearing some kind of strange clothes.

MON: I want to move from your physical location in space to another physical location, but in this time period. Move now to: 46.45 north 353.22 east. Move in this time to: 46.45 north 353.22 east.

SUB: Deep inside of a cavern, not a cavern, more like canyon. Um, I'm looking up, up the sides of a steep wall that seem to go on forever. And there's like ah . . . a structure with a . . . it's like the wall of the canyon itself has been carved. Again I'm getting a very large structures, no . . . ah . . . no intricacies, huge sections of smooth stone.

MON: Do the structures have insides and outsides?

SUB: Yes, they're very, it's like a rabbit warren, corners of rooms, they're really huge, I don't, feel like I'm standing in one it's just really huge. Perception is that the ceiling is very high, walls very wide.

MON: Yes that would be correct. All right, I'd like to move now to another location nearby. All right, move from this point in this time to: 45.86 north 354.1 east 45.86 north 354.1 east.

SUB: They have a ah . . . appears to be the end of a very large road and there's a . . . marker thing that's very large, keep getting Washington Monument overlay, it's like an . . . obelisk.

MON: All right. From this point then, let us move to another point. Move now to: 35.26 north 213.24 east. Move in this time to: 35.26 north 213.24 east.

SUB: It's like I'm in the middle of a huge circular basin . . . of the range mountains by almost all the way around . . . very ragged, ragged mountains, very tall. Basin's very, very, very large. Scale seems to be off or something, it's just really big, everything's big.

MON: I understand the problem, just continue.

SUB: See just a right angle corner to something but that's all, I don't see anything else.

MON: Okay. Then let's move into a little different place, very close. Move from the point you are now, in this time, to: 34.6 north 213.09 east. Move now in this time to: 34.6 north 213.09 east.

SUB: The cluster of squares up and down. Um . . . it's like you want to make them square anyway. They're almost flush with the ground and it's like they're connected. . . . Something very white or reflects light.

MON: What's your position of observation as you look at this thing that reflects light?

SUB: I'm amid ah . . . oblique left angle, sun is ah . . . sun is weird.

MON: Look back down at the ground now, and we're going to move just a little bit from this place, just a little bit from this place. 34.57 north 212.22 east. Very close by. Now, move over now to: 34.57 north 212.22 east.

SUB: It's like I can just perceive ah . . . like a radiating pattern of some kind. It's like some really . . . ah . . . strange intersecting kind of roads that are dug into valleys, you know, where a road is just a little below the edge.

MON: Tell me about the shapes of these things.

SUB: They're like real neat channels cut, they're very deep, it's like the road went down.

MON: Okay. Now I have, I notice electrically you're nulled out a little bit and I want you to stay deep and recapture your focus here.

SUB: It's really tough, it's seems like it's just always very sporadic.

MON: I realize that, it's very important that you maintain your focus. I have a movement exercise again for you and this is some considerable distance away, so holding the focus in time, remember the focus in time that you had before and moving now to: 15 degrees north 198 degrees east. Take some time and get back deep.

SUB: See the . . . um, intersecting ah . . . whatever these are, are aqueduct type things . . . these . . . rounded bottom carved channels, like road beds. See ah . . . see pointed tops of something on the horizon. Even the horizon looks funny and weird, it's like ah . . . different . . . misty, like it's really far away . . . very vague.

MON: Okay. Another movement now to: 80 degrees south, 80 degrees south 64 degrees east, 64 degrees east. Move now in this time to: 80 degrees south 64 degrees east.

SUB: See pyramids. . . . Can't tell if it's overlay or not 'cause they're different.

MON: Okay. Do these pyramids have insides and outsides?

SUB: Um-hum, got both, and they're huge. It's really, ah . . . it's an interesting perception I'm getting.

MON: (I think that he's losing his ability to move accurately, but he is attracted to things that are interesting, so we're going to go with his own, we're going to let him go ahead and explore what seems to be interesting to him rather than move on the targets indicated here.)

SUB: It's filtered from storms or something.

MON: Say that again, SUB.

SUB: They're like shelters from storms.

MON: These structures you're seeing?

SUB: Yes. They're designed for that.

MON: All right. Go inside one of these and find some activity to tell me about.

SUB: Different chambers, . . . but they're almost stripped of any kind of . . . furnishings or anything, it's like ah . . . strictly functional place for sleeping or that's not a good word, hibernations, some form, I can't, I get real raw inputs, storms, savage storm, and sleeping through storms.

MON: Tell me about the ones who sleep through the storms.

SUB: Ah . . . very . . . tall again, very large . . . people, but they're thin, they look thin because of their height and they dress like in, oh hell, it's like a real light silk, but it's not flowing type of clothing, it's like cut to fit.

MON: Move close to one of them and ask them to tell you about themselves.

SUB: They're ancient people. They're ah . . . they're dying, it's past their time or age.

MON: Tell me about this.

SUB: They're very philosophic about it. They're looking for ah . . . a way to survive and they just can't.

SUB: Can't seem to get their way out, they can't seem to find their way out . . . so they're hanging on while they look or wait for something to return or something coming with the answer.

MON: What is it they're waiting for?

SUB: They're ah . . . evidently was a . . . a group or a party of them that went to find ah . . . new place to live. It's like I'm getting all kinds of overwhelming input of the . . . corruption of their environment. It's failing very rapidly and this group went somewhere, like a long way to find another place to live.

MON: What was the cause of the atmospheric disturbance or the environment disturbance?

SUB: I see a picture of a, picture of like a, oh hell, it's almost a warp in a, oh god, this is difficult. It's like going, let's see.

MON: The raw data?

SUB: Oh, I get a globe . . . ah . . . it's like a globe that goes through a comet's tail or . . . it's through a river of something, but it's all very cosmic. It's like space pictures.

MON: All right, now before you leave this individual, ask him if there is any way that you, ask him if he knows who you are and is there any way you can help him in his present predicament?

SUB: All I get is that they must just wait. Doesn't know who I am. Think he perceives I'm a hallucination or something.

MON: Okay, when the others left, these people are waiting, when the others left, how did they go?

SUB: Get an impression of ah . . . don't know what the hell it is. It looks like the inside of a larger boat. Very rounded walls and shiny metal.

MON: Go along with them on their journey and find out where it is they go.

SUB: Impression of a really crazy place with volcanos and gas pockets and strange plants, very volatile place, it's very much like going from the frying pan into the fire. Difference is there seems to be a lot of vegetation where the other place did not have it. And different kind of storm.

MON: All right it's time to come back now to the sound of my voice into present time to right now the 22nd of May 1984, the sound of my voice. Move now back to the room, back to the sound of my voice, back further now to the sound of my voice on the 22nd of May 1984.

There's not a shred of doubt that the images the remote-viewing session produced were hellish in the extreme. Disturbing references to the "corruption of their environment," to "atmospheric disturbance," to "a very volatile place," to "severe, severe clouds," to a "geologic problem," and to a race of terrified, giant Martians doing their very best to survive in massive "shelters" deep below the planet—collectively, this provokes an end-of-the-world-type scenario of a truly horrific kind. It's an image very much like that presented in a 2009 movie starring Nicolas Cage, *Knowing*, in which advanced aliens—or angels (it's never made entirely clear)—whisk away a large number of children who are destined to survive as the Earth is completely fried to a crisp from the effects of a massive solar flare. Not just the human race, but everything on Earth is exterminated. A thriving world is now a huge, scorched, ball of rock. In a strange and alternative way, *Knowing* may give us some degree of insight—albeit fictional insight—into what the Martians went through in their final days in an era long, long gone.

What's particularly intriguing about all of this is *why*, exactly, the CIA had a pressing need to remote-view Mars in the first place. The agency has most definitely not shared that important nugget. It must be noted that in the very same way that the CIA had a fascination for the remote viewing of Mars, it also had—and still has—a deep interest in *other* mysteries of the distant past. For example, the CIA has a large file on the legend of Noah's Ark, a small portion of which has been declassified under the terms of the Freedom of Information Act. It addresses the theory that the mighty ship was actually an extraterrestrial spacecraft, one that was created by highly advanced aliens who did their utmost to try and save certain portions of the human race when a catastrophic flood overwhelmed our world in the distant past.

On similar territory, there is the matter of H. P. Albarelli. He's the author of a 2009 book titled *A Terrible Mistake*. It focuses on Cold War controversies and the CIA. He writes: "Several heavily redacted CIA documents reveal a keen interest in the Ark of the Covenant," in the "rock at Horeb," and in "Solomon's Temple." He also states that the CIA had an interest in the "peculiar apparatus reportedly witnessed by Ezekiel." Albarelli presents those words, you'll note, in quotation marks, suggesting they are extracted directly from CIA papers.[6]

What all of this tells us is that certain factions of the CIA don't just have an interest in the mysteries of the past, but that they also know—or, more likely, suspect—a great deal about it, too. The

CIA's quest to learn what happened to the Martians—in an ancient era—was without doubt, however, the most significant of all such semi-related projects. Let's hope that one day the agency will finally let *us* know what *it* knows of the Martian race: dead or alive, still on Mars, or hiding out on our very own world—the latter scenario being a sensational subject that we'll get to in a later chapter.

CHAPTER 9

"Strange Geometric Ground Markings and Symbols"

||||||||||||||||||||||||||||||||||||||

Of one thing we can be certain when it comes to the matter of the CIA's 1984 remote viewing of Mars: The entire story is undeniably and wholly incredible. We are, after all, talking about an ancient Martian race, giant-sized humanoids, ancient pyramids, a civilization teetering on the very edge of almost certain, irreversible destruction, and much more. Taking all of these astonishing revelations into consideration, one would imagine that the CIA would have hastily continued its research to try and solve the mystery of Mars's history—and that of its inhabitants, too. At first glance, however, it appears that the CIA's May 22, 1984, operation to try to learn what it could about Mars was very much a solitary, single affair. After all, what we have is a single experiment that appears not to have been expanded on to any degree in the slightest. It has all the signs of being nothing more than a single, quirky, off-the-wall experiment that was shelved—and that was all. Nothing else whatsoever on the events of 1984 has ever surfaced under the terms of the Freedom of Information Act.

On top of that, no further documentation adding to and expanding on the story in later years has come to light, either. But, is that really the beginning and the end of the matter? Maybe not. It's important to note there is evidence to suggest that the US Intelligence Community has—in a highly secretive fashion—used remote viewing as a means to try and secure amazing secrets concerning our nearest neighbor, the Moon, as we shall see now. It's a sensational story that also ties in with that of the Martians. It is thanks to the revelations of one Ingo Swann that we know this.

Swann was a highly skilled remote viewer who died in 2013 at the age of seventy-nine, and someone who was one of the key figures in the creation of the US Army's psychic-spying project called "Stargate." It operated out of Fort Meade, Maryland, and under the auspices of the Defense Intelligence Agency. And, in an all-enveloping cloak of secrecy, it scarcely needs stressing. Swann was used to working in an environment that was shrouded in secrecy and that impacted national security. As for what Swann learned about anomalies on the Moon and Mars, it all went down in 1975, specifically in the early days of February. As will soon become apparent, the situation Swann found himself in eerily mirrored the CIA's experiment of 1984—something that leads me to believe the two operations were actually parts of a much bigger, interconnected program, although admittedly I cannot prove that. I doubt that anyone else outside of the Intelligence Community can either.

It must be said that Swann was very careful and cagey—even to the point of exhibiting outright paranoia when he first decided to tell the story—about how he described the whole, strange situation. Names were changed. Locations were obfuscated. And shadowy characters dominated the whole thing. It all began when Swann received a phone call, at home, late one night in 1975.[1] The call came from a man who Swann said was a powerful and influential character who worked in the heart of American intelligence in the D.C. area, which was and still is *the* veritable hub of espionage and secrecy. The caller did not identify himself; he did, however, advise Swann that another man would soon be contacting him—if, that was, Swann was up for a certain, mind-blowing challenge.

Taking into consideration the fact that the man on the line was acting in a decidedly Machiavellian fashion, Swann told his source that it was difficult for him to know what to say or do, given that almost nothing was being made clear to him. The man told Swann not to worry: All would quickly be explained. And, Swann's caller said, it all revolved around the world of remote viewing. Well, that was enough for Swann—for then, at least. About a month after the curious call there was finally a development. As Swann guessed— and as we should, too—the contact came via a phone call. Things were about to get weird. Let me correct that: They were about to get *very* weird.

The man on the end of the phone—whom Swann was able to confirm was not the same man who had contacted him a few weeks earlier; the accent gave that much away—told, rather than asked, Swann to make his way to the National Museum of Natural History. Both anxious and excited, Swann did exactly as he was ordered. A cab driver soon had Swann outside the doors of the historic museum. Swann wasn't sure what he should do next. It didn't matter, however, as someone soon *told* him what to do: a well-built man, with a cropped, military-type haircut, and dressed in a suit, and whom Swann estimated was in his mid-thirties or thereabouts. The man told Swann that the journey was not over. In fact, far from it. From this point onward, the whole thing took on the tone of a high-octane spy movie.

Swann was directed to a nearby black car, which had a small flag affixed to the hood. The mysterious character motioned Swann to get in the back of the car. He did so. The mysterious man got in the front passenger seat, while an elderly, white-haired man drove away from the museum, saying absolutely nothing for the entirety of the journey. A sudden, freezing chill went through Swann's veins when the man turned his head toward Swann and slowly took something out of his jacket pocket. Swann's first thought was that the man was about to point a gun in his (Swann's) direction. Thankfully, the man did not. What he did do, however, was reveal nothing less than a blindfold: a black piece of material that was held in place by a piece of elastic that fitted tightly around Swann's ears and head. Swann was told not to worry; it was just a simple precaution and nothing else. The man explained to Swann that the blindfold was essential because he would soon be taken to an airstrip and, on arrival, would then be taken by helicopter to meet the man who had engineered the weird affair that Swann was now so deeply—and clearly irreversibly—enmeshed in. The drive was relatively short: Swann estimated less than an hour.

In somewhat clumsy fashion, Swann—still blindfolded—was then taken from the car and onto a helicopter. As was the drive, the helicopter flight was short, maybe only about twenty minutes at the very most, Swann recollected. On landing, Swann was then directed to a building by what he suspected was yet another man, one who put his hands on Swann's shoulders and directed him in which way to move, all in complete and utter silence. Talk about surreal! Finally, Swann was suddenly stopped in his tracks and guided through a door. When the door was closed, the blindfold was removed and Swann could now see—in a somewhat blurry fashion for a few moments—he was in a windowless room. As for the location, Swann had no idea at all. A woman of about forty, and dressed in a business suit, motioned Swann into an elevator. Swann continued to do as he was told. The descent, said Swann, went on for what he said was a surprisingly long time and clearly to what was a deep depth. That much was made clear by the fact that Swann's ears popped on a couple of occasions. When the doors opened, Swann finally met with the man who was responsible for the strange events of the day: a certain Mr. Axelrod.

By Swann's own admission—and largely due to the shenanigans of the previous few hours—he doubted that Axelrod was the man's real name. From a source that he declined to reveal, however, Swann claimed to have later learned that the man's real name was Raymond Wallis; although, for reasons that aren't clear, Swann suspected that "Raymond" was the man's middle name. True or not, it all presents us with yet another layer to the mystery. It was clear to Swann that the man using the name of Axelrod knew all about his work in the field of remote viewing and that he (Axelrod) wanted Swann to do some clandestine work on his behalf of certain figures in the Intelligence Community. Swann had already deduced that much. After all, his time spent in the world of government secrecy had led him to suspect that whatever Axelrod wanted, it surely impacted national security.

Time seemed to be of the essence and Axelrod, clearly a man who only spoke when he needed to, told Swann that he was impressed by his work in the arena of psychic spying. In fact, Axelrod made sure that Swann could see a dossier on a table that had Swann's name on the cover page. Clearly, this was a carefully planned and placed mind game designed to let Swann know he was being watched—and that he probably had been for a very long time. Axelrod was the kind of guy who knew what it took to get the wheels in motion: money—and a considerable amount of it was on offer, too. The combination of a new, secret program to work on, and a considerable fee for his time, led Swann to say *yes!* to the offer, even before he had any meaningful understanding of what it was really all about.

Axelrod, apparently satisfied that things were progressing nicely and incredibly quickly, got right to the heart of things. He had an eye-opening question for Swann: "What do you know about the Moon?" Now, it suddenly became all too clear: Axelrod, or maybe someone that he was working for, wanted a skilled remote viewer to take a look at a certain portion of the lunar surface. But, what, exactly, was that same portion of the Moon? Swann would soon come to know.

Over the course of several weeks, Swann was taken back and forth to that mysterious underground facility, where he would focus his mind skills on what he could find on one particular section of the far side of the Moon, which, Swann was told, was what Axelrod was particularly interested in. The first thing that Swann was able to "see," using the tried and tested processes that apply to remote viewing, was a huge obelisk. From the way Swann described it, the towering monument was most definitely not a natural creation; it was clearly fashioned by intelligent entities. But, who? And when? And why? Those were Axelrod's primary questions. And, recall that an

Martians under the far side of the Moon. (NASA)

obelisk was seen by the remote viewer who worked on the Mars program in 1984. As the weeks progressed things became stranger and stranger—if such a thing were even possible. Three additional obelisks were said to have been found, although none were anywhere near the size of the first one that Swann had successfully found. He also "saw" huge tube-like creations embedded in the lunar landscape; they were tubes that resembled, in Swann's own words, gigantic "soda straws." Interestingly, they were very much akin to those huge "straws" that were found on Mars years later, as we saw in an earlier chapter. And, on top of that, we'll also soon learn all about obelisks similar to the ones Swann described, which were discovered on both Mars and one of its two moons, Phobos. It was Swann's deep suspicion that the tubes were a means of transport.

Most amazing of all—and controversial, too—Swann's mind became flooded by images of huge, incredibly complex machines that were busily mining the airless landscape of our very own Moon. Swann quickly concluded that extraterrestrials were secretly building below-surface facilities—of a sprawling, mega-size—on our Moon. Disturbingly, and during this very same time frame, Swann developed a "vibe" that there was something very worrying about all of this: that there was something deeply sinister about what was afoot on the Moon—and possibly deep below it. Yes, the most fantastic part of the story was still yet to come.

On one memorable occasion, Swann managed to focus his mind on those below-surface areas of the Moon and saw something that was clearly both fantastic and terrifying: near-human-like creatures busily working on all kinds of bizarre operations and experiments—most of them concentrated on burrowing further and further into the Moon. The three things that really stood out were (1) the humanoid aliens were giant-sized: incredibly, somewhere in the region of 9 or 10 feet in height, (2) they were all naked—at all times, and (3) they were not natives of the Moon. They were the last remnants of an ancient Martian civilization that was struggling to stay alive, after a planetary apocalypse all but destroyed them long ago.[2] They were doing their absolute utmost to keep extinction at bay by making portions of our Moon their new world of the underground variety. This made Swann wonder, and worry, if the Moon was being turned into a kind of massive staging post that, one day, just might allow the Martians to launch a planetary assault on our world and quickly claim it as theirs. Those issues—of the Martians being Goliath-like in size, and of the ragged survivors of a Martian apocalypse frantically heading to our Moon (and, possibly, the Earth, too), as a means to try and ensure their survival—will surface again. And, on several occasions, no less.

There was one more thing left to discuss. It was something that Axelrod had kept until the absolute final moment—almost surely deliberately and for a very good reason. When Swann explained all that he had seen, Axelrod's face took on a grim appearance. He explained to Swann that there were other remote viewers aboard, whose names Swann was not told, which puzzled him. This led him to conclude there were other groups in the Pentagon, all also doing work that revolved around remote viewing and with connections to both the Moon and Mars. In a somewhat-uneasy fashion, Axelrod told Swann that some of the other remote viewers who had been tasked to look at the Moon came to realize, and to their absolute horror, that those huge aliens were able to know when they were being watched by the Pentagon's psychic spies. In the same way that *we* were watching *them, they* were watching *us.*

It got worse: Axelrod said that the aliens now almost certainly had the ability to find Swann, at any time and any location—and maybe even terminate him and the other remote viewers if it was deemed necessary. Swann, then, had become a marked man. It has to be said that this was definitely not what Swann had signed up for.

No wonder Axelrod kept the very worst until the very last, thought Swann, by then both worried and angered. Had he known about the possibility of being wiped out by Martians who were on the verge of extinction, Swann protested to Axelrod, he would probably not have gotten involved. Of course, Axelrod astutely knew that, which is why he chose not to mention it until matters were all over. Nevertheless, Swann agreed to periodically continue to do work for Axelrod, most of which failed to uncover anything further, beyond that someone was fiercely and frantically hollowing parts of our Moon—and it wasn't us. It was a band of Martians, seeking ways to avoid extinction.

The Mars-Moon connection. (NASA)

One final thing: On one particular occasion Axelrod casually asked Swann if he had heard of a man named George Leonard. At the time, Swann had not heard of Leonard. Axelrod told Swann, on a date in 1975 that Swann could not recall when he went public with the story, that Leonard was working on a book that was, said Axelrod, "of interest to us."[3] Not only that: The book would focus, in part, on some of the things that had concerned Axelrod and his team in 1975. The implication was that someone "on the inside" was secretly and constantly monitoring Leonard's research and writing. Swann was amazed when, in 1976, he was provided— by Axelrod—with a then-newly surfaced copy of Leonard's book, *Somebody Else Is on the Moon*, which was published by David McKay. Notably, the back cover of the book details for the reader some of the highlights of the story: "Immense mechanical rigs, some over a mile"; "Strange geometric ground markings and symbols"; "constructions several times higher than anything built on Earth"; and "Somebody is doing something on our Moon—and doing it right now, on a stunningly massive scale!"[4] To Swann, this sounded very much like what Axelrod knew back in 1975. And it most certainly mirrored what Swann had found for Axelrod.

So, what do we have here? Let us take a close and a careful look. At some point midway through the 1970s someone in the US government, military, intelligence community, or a swirling combination of all three, felt it was deeply important—pressing, even—to take a close look at the far side of the Moon, which, by the way, never shows itself to us. It cannot be seen at all unless one takes a flyby of that same elusive side. The account of Ingo Swann, and the incredible degree of security that surrounded his involvement in what was clearly a top-secret project, suggests strongly that national security was at the forefront of officialdom's concern. We also know that at the same time the US government was deeply worried about the things that were occurring on that elusive-to-the-eye portion of the Moon and in relation to ancient Martians, so was George Leonard. And, incredibly, we also know that Ingo Swann concluded—as a result of his remote-viewing skills—that the creatures that apparently lived deep below the surface of our Moon were Martians, feverishly seeking ways to keep themselves alive. Swann said he saw an obelisk on the Moon, and we know now that one of the many anomalies found on the surface of Mars's moon Phobos has its very own obelisk, as will soon be made clear.

What all of this demonstrates is the following: The CIA's quest of May 22, 1984, for the truth of what happened to Mars and to its Martian race does *not* stand alone, after all. It was *not* a one-off experiment. Back in 1975 the mysterious Mr. Axelrod and Co. knew of an ancient Martian connection to the Moon, thanks to the supernatural skills of Ingo Swann. In light of all this, it's not at all unfeasible or impossible that numerous additional studies of the Martian history—and of the connection to the Moon—were undertaken between 1975 and 1984 by the US government. Maybe, the investigations are still afoot—and, who knows, possibly with NASA left completely out of the loop.

One final thing to ponder: The CIA's remote-viewing operation took place on May 22, 1984. That was just two months before the Case for Mars II Conference took place, in Boulder, Colorado. It was designed to determine the then-current state of research into the matter of the Face on Mars. One of the presentations came from Richard Hoagland. It was titled "Preliminary report of the Independent Mars Investigation Team: New evidence of Prior Habitation?"[5]

That Hoagland was busy digging into the Face on Mars issue in much of 1984—and that the CIA was digging into Mars in 1984, too—makes me wonder if the CIA was keeping a careful and quiet watch on what Hoagland was up to, research-wise.

CHAPTER 10

"Unusual Images Were Radioed back to Earth"

‖‖‖‖‖‖‖‖‖‖‖‖‖‖‖‖‖‖‖‖‖‖‖‖‖‖‖‖‖‖‖‖‖‖‖‖

"After a seventeen-year gap since its last mission to the Red Planet, the United States launched *Mars Observer* on September 25, 1992. The spacecraft was based on a commercial Earth-orbiting communications satellite that had been converted into an orbiter for Mars. The payload of science instruments was designed to study the geology, geophysics, and climate of Mars," says NASA. Things didn't quite work out right, however, as NASA explains: "The mission ended with disappointment on August 22, 1993, when contact was lost with the spacecraft shortly before it was to enter orbit around Mars. Science instruments from *Mars Observer* were reflown on two other orbiters, Mars Global Surveyor and 2001 Mars Odyssey."[1]

AmericaSpace details what happened next:

Contact with *Mars Observer* was abruptly lost, for no obvious reason, and repeated attempts to communicate with the spacecraft proved fruitless. It was never heard from again. Five months later, in January 1994, an investigation panel from the Naval Research Laboratory (NRL) concluded that the most likely cause of the spacecraft's disappearance was a ruptured fuel pressurization tank in its main propulsion system. Hypergolic monomethyl hydrazine may have leaked past valves during the 11-month journey to Mars and inadvertently mixed with nitrogen tetroxide. The leaking fuel may have induced an extremely high spin-rate and likely damaged critical components aboard Mars Observer itself.[2]

Perhaps inevitably, not everyone bought into the theory/conclusion that the *Mars Observer* mission came to a shuddering end as a result of nothing stranger than problems with a fuel pressurization tank, and particularly those in the field of conspiracy-theorizing. In their 2007 book, *Dark Mission: The Secret History of NASA*, Richard Hoagland and Mike Bara said that NASA ". . . had inexplicably ordered *Mars Observer* to shut off its primary data stream prior to executing a key pre-orbital burn. . . . Because NASA had violated the first rule of space travel—you never turn off the radio—no cause for the probe's loss was ever satisfactorily determined."[3]

James Oberg, a well-known skeptic when it comes to the matter of NASA and conspiracy theories, said:

Dark Mission portrays the failure of the *Mars Observer* probe in 1993 as a deliberate act by NASA to prevent the publication of its expected photographs of artificial Martian ruins. But the description of the events is inconsistent with well-documented accounts, reports non-existent events, and omits well-known explanations for important features of the probe's flight plan. All of this can be easily confirmed through Internet searches.[4]

So, why was that radio turned off? NASA provided its answer in its 313-page report, "Failure Investigation Board Report." The relevant section of the report provides us with the following:

In accordance with the mission's published flight rules, the transmitter on the spacecraft had been turned off during the propellant-tank Pressurization Sequence on 21 August. . . . To protect the spacecraft radio frequency transmitter from damage during the Pressurization Sequence (albeit a very low probability), the software included a command to turn off the Mars Observer transponder and radio frequency (RF) telemetry power amplifier for a period of ten minutes. This was a standard procedure that had been implemented several times earlier during the mission.[5]

Conspiracy theorists doubted the official story, and for one particular reason they saw as proof there was far more to the failure of the *Mars Observer* than met the eye: Mark Strauss, of the *Smithsonian* website, explained why such suspicions developed:

During the 1990s, NASA lost three spacecraft destined for the Red Planet: the *Mars Observer* (which, in 1993, terminated communication just three days before entering orbit); the *Mars Polar Lander* (which, in 1999, is believed to have crashed during its descent to the Martian surface); and the *Mars Climate Orbiter* (which, in 1999, burned up in Mars' upper atmosphere). Conspiracy theorists claimed that either aliens had destroyed the spacecraft or that NASA had destroyed its own probes to cover-up evidence of an extraterrestrial civilization.[6]

That's right: No less than three missions ended badly in the Nineties. It should be noted that space-flight—manned or not—is incredibly dicey at the very best of times, and one should not jump to the conclusion that three such failures, all revolving around missions to Mars, were the result of anything mysterious or conspiracy-driven. Nevertheless, the theory that NASA deliberately sabotaged the three missions, specifically to try to prevent evidence surfacing of the remains of an ancient Martian race civilization, still exists. It's fair to say that it positively thrives. As evidence of this, and as James Oberg noted, the Hoagland-Bara book, *Dark Mission*, "came within one tick mark of making it onto the *New York Times* bestsellers list for paperback nonfiction."[7] I probably don't need to explain that Oberg was not happy with that news.

A world of lost and hidden treasures. (NASA)

The theory that NASA has sabotaged a number of missions to Mars was resurrected in October 2016. Writing for *Tech Times,* Rhodi Lee said:

> The *Schiaparelli* Mars lander lost contact with the European Space Agency (ESA) controllers on Oct. 19, about one minute before the European probe was about to land on the surface of the Red Planet. Images taken by NASA's Mars Reconnaissance Orbiter showed what appear as the parachute and crash site of the European lander suggesting that Schiaparelli may not have survived in its attempt to touch down on Mars. It is suspected that the entry,

descent and landing sequence of the probe did not go as planned, with the failure believed to be associated with the parachute's timing and the retro-rockets driving the lander into the surface of Mars. The lander appears to have suffered from a computer glitch.[8]

UFO investigator Scott Waring suggested that NASA shot down the ESA probe as a means to try to ensure that when an announcement is made to the effect that, yes, there is life on Mars, it will be NASA, and not the ESA, who will make the incredible announcement. There's another reason why this conspiracy theory persists: It's not just NASA and the ESA that have had problems with their probes to Mars. Russia has had its own share of problems with its Mars probes.

"*Phobos 2* launched toward Mars on July 12, 1988. It successfully went into orbit on January 29, 1989, and began sending back preliminary data," says the Planetary Society. They add:

Several sets of images were returned on February 21 and 28, as *Phobos 2* orbited Mars in a path coplanar with the orbit of the moon, but 300 kilometers (180 miles) farther from Mars. Because of the larger distance from Mars, *Phobos 2* traveled more slowly than the moon, so images were only possible during "flybys" that occurred once a week. Next, the spacecraft adjusted its orbit so that its orbital period was the same as that of Phobos, allowing it to remain stationed within 500 kilometers (300 miles) of the moon. Five more sets of images were acquired on March 25. Then, on March 27, just before the spacecraft was to move within 50 meters of Phobos and deploy the two landers, the spacecraft's onboard computer malfunctioned and the mission was lost.[9]

What happened next had UFO researchers in states of overwhelming excitement. The *Keys of Enoch* detail the story:

One of the last images relayed to earth in detail by the *Phobos II* camera before data transmission was lost was an enormous elliptical shadow on the surface of Mars—cigar-shaped, and an estimated 25–27 kilometers (approximately 16 miles) in length. The size of this object ruled out the possibility that it was a reflection of the *Phobos* spacecraft itself. Because of its position, its symmetrical shape, its size and its movement, no features on the surface of Mars in the area in front of the probe, nor the satellite moons of Phobos and Deimos, nor the *Phobos II* spacecraft itself could account for this shadow pattern occurring in the very last frames of data successfully transmitted to earth.[10]

The Dark Star website adds further fuel to the flames:

According to Boris Bolitsky, science correspondent for *Radio Moscow*, just before radio contact was lost with *Phobos 2*, several unusual images were radioed back to Earth, described by the Russian as "Quite remarkable features." A report taken from *New Scientist* of 8 April 1989, described the following: "The features are either on the Martian surface or in the lower atmosphere. The features are between 20 and 25 kilometers wide and do not resemble any known geological formation. They are spindle-shaped and proving to be intriguing and puzzling."[11]

The matter was never resolved.

CHAPTER 11

"The Image Is
So Striking"

||||||||||||||||||||||||||||||||||||||

"I'm now convinced that Mars is inhabited by a race of demented landscape gardeners."[1] Those were the eye-opening, amusing words of the late Arthur C. Clarke, one of the world's most revered of all science-fiction writers, who was born in 1917 and who died at the ripe old age of ninety. Clarke's words about those demented landscape gardeners were largely made in jest, but they served to make an important and astonishing point: Clarke came to believe that evidence had been found strongly suggesting that there was vegetation on the Red Planet. He even suggested—tentatively, admittedly—that Mars just might be *teeming* with trees, bushes, and plants. And if there was vegetation, then there just might be other kinds of life on Mars, too. Possibly even intelligent life. And what was it that led Clarke to come to such a controversial conclusion? Nothing less than NASA's very own priceless photos of Mars's surface, that's what. But, let's not get too far ahead of ourselves. How did such a controversy begin?

For Arthur C. Clarke, matters began in his teenage years. That's when he developed a fascination—which lasted throughout his life—with the mysteries of outer space. Such was the level of that fascination that Clarke became a member of the British Interplanetary Society (BIS) before he was twenty. During the Second World War, Clarke worked in the field of radar in the British Royal Air Force, honing his interest in science and technology. With the war finally over in 1945, and as a result of his growing enthusiasm for worlds beyond ours, Clarke accepted the position of chairman of the BIS. He oversaw it for two years. Clarke then took a break, coming back to run the organization from 1951 to 1953.

Clarke became not only a well-respected figure as an inventor, a writer, and an explorer, he was also someone who had an overriding passion for science fiction. In this field, particularly, Clarke absolutely thrived. For example, there was his short story "The Sentinel," which was written in 1948 and published three years later. It served as the direct innovation for Stanley Kubrick's classic 1968 movie, *2001: A Space Odyssey*. In 2015, the SyFy Channel broadcast a three-hour-long series based on Clarke's 1953 novel, *Childhood's End*. Clarke also turned his talents to television, presenting in the 1980s *Arthur C. Clarke's Mysterious World; Arthur C. Clarke's World of Strange Powers;* and *Arthur C. Clarke's Mysterious Universe*. He died on March 19, 2008. All of this brings us back to that controversial statement of the "demented" kind. It would never have been uttered had it not been for certain, spectacular photos that came back from NASA's Mars Global Surveyor.

As NASA noted in 1997: "Launched November 7, 1996, *Mars Global Surveyor* became the first successful mission to the red planet in two decades. After a year and a half spent trimming its orbit from a looping ellipse to a circular track around the planet, the spacecraft began its prime mapping mission in March 1999. It has continued to observe the planet from a low-altitude, nearly polar orbit ever since."[2] Not only that, NASA added that Mars has "very repeatable weather patterns" and that a "panoply of high-resolution images from the *Mars Global Surveyor* has documented gullies and debris flows suggesting that occasional sources of liquid water, similar to an aquifer, were once present at or near the surface of the planet."[3]

That same panoply revealed something else—something amazing, if the data was not being misinterpreted. Tucked away among a wealth of less-controversial images from the *Mars Global Surveyor* were a number of astounding images that appeared to show nothing less than vast areas of vegetation—trees, even. They looked eerily like what on Earth are termed *banyan trees*. They are, essentially, trees that grow and thrive by living on other trees. It wasn't long before the media—and Arthur C. Clarke—caught wind of these extraordinary photographs and what they seemed to show. He wasted no time at all in giving his opinion on this new development on Mars. Clarke made no bones about it when he said that the collective images were "so striking that there is no need to say anything about it."[4] Clarke was also excited by the fact that Mars's

The saga of the Martian trees. (NASA)

very own banyan trees appeared to alter in appearance, according to the seasons on Mars. On this particular point, Clarke could not forget NASA's words that Mars had "very repeatable weather patterns" when he said, "Something is actually moving and changing with the seasons."[5]

The Mother Nature Network suggested a different, and far less exciting, possibility for that which had Clarke in a state of excitement: "Few serious scientists agreed [with Clarke], however, and a later study presented a number of competing, more likely theories. For instance, dark basaltic sand pushed to the surface of sand dunes by sun-heated solid carbon dioxide, or dry ice, sublimating directly into vapor creates the illusion of a tree. This occurs seasonally, as the sun heats the Martian surface during that planet's springtime."[6]

Candy Hansen, a key figure on the *Mars Reconnaissance Orbiter* program, said, in forthright fashion: "To date, there is no firm evidence of any type of Martian biology, past or present, plant or otherwise. In the Martian spring, the sun warms the ice, causing it to sublimate directly into vapor, and the resulting gas dislodges surrounding dust and sand particles. What we think is happening is that the dark sand is sliding down the bright frosted portion of the dune."[7]

Popular Science got involved in the controversy, too. They put a few questions to Clarke, who was certainly the most visible of those who believed that Mars just might have trees dotted around its landscape. They put the following to Clarke: "What makes you so confident there is life on Mars?"

Clarke replied: "The image is so striking that there is no need to say anything about it—it's obviously vegetation to any unbiased eye."

Popular Science then turned their attentions to something else: "What about animal life?"

Clarke had an answer: "If there is vegetation, it seems probable there are other life-forms as well." When it was put to Clarke that not many of his peers concurred with him, he shot back: "They are right to be cautious; we still don't have one hundred percent proof. I think it's in the high nineties." When he was asked why he was so fired up, Clarke came straight to the point: "Because nothing could be more important than the discovery of other life-forms. It's getting lonely down here."[8]

The banyan trees are not the only phenomenon on Mars that have some researchers of Martian enigmas convinced that plant-life exists on Mars. There's also the curious matter of the planet's "Dalmatian spots," as they became known. They are strange-looking phenomena that have many convinced that this is yet further evidence of life on Mars. On June 10, 2003, and in response to photographs of the spots taken by NASA's 2001 *Mars Odyssey* craft at its polar regions, the agency's staff made it clear they weren't buying into the undeniably exotic theory: "The polar regions of Mars have surfaces than can show dark spots on a brighter background. These surfaces are informally called 'Dalmatian terrain' because of their appearance." NASA continued: "Elsewhere, defrosting dunes have shown a similar spotted pattern. Perhaps this 'Dalmatian terrain' is a distinctive pattern that forms over all defrosting patches of sand."[9]

I spoke with Mars researcher Mac Tonnies about NASA's theory concerning the Dalmatian spots. He was far from being impressed by the less-exotic theories that NASA had presented to the public and the media: "The universally dark coloration of these anomalies suggests chlorophyll, the pigment that allows plants to convert carbon dioxide into oxygen. Indeed, Russian astronomers claim to have detected organic pigment in Mars's atmosphere, presumably from a planetary ecology."[10]

Now, let's address the matters of Mars's giant-sized spiders.

Yes, you did read that right. But, no, we're not talking about something like a wild plot from a 1950s sci-fi-driven *Godzilla*-type movie. We're actually talking about something far more exciting. It's a story that dates back to 1999, which was when the *Mars Global*

Surveyor photographed some seriously strange "things" on the surface of Mars. "Spidery" is the best way to describe them. One person, more than any other, who took up the challenge to solve the mystery is Greg Orme. Someone who has a deep passion for the mysteries of Mars, Orme began his research into Martian anomalies in 1994. Since then, he has written a book, *Why We Must Go to Mars: The King's Valley,* and a paper on those mysterious spiders for the journal of the British Interplanetary Society. Orme's voluminous collection of photos of Martian anomalies is highly impressive. Mac Tonnies was excited by Orme's work and went on to describe the "Black Spiders," as they became known, as resembling "nerve ganglia" or "ground-hugging trees" within "a macabre forest." Such descriptions are undeniably appropriate.

Predictably, NASA has a very different view on the spiders of Mars:

> These aren't actual spiders. Called "araneiform terrain," they are spider-like radiating mounds that form when carbon dioxide ice below the surface heats up and releases. This is an active seasonal process not seen on Earth. Like dry ice on Earth, the carbon dioxide ice on Mars sublimates as it warms (changes from solid to gas) and the gas becomes trapped below the surface. Over time, the trapped carbon dioxide gas builds in pressure and is eventually strong enough to break through the ice as a jet that erupts dust. The gas is released into the atmosphere and darker dust may be deposited around the vent or transported by winds to produce streaks. The loss of the sublimated carbon dioxide leaves behind these spider-like features etched into the surface.[11]

It should be noted that the controversy surrounding the spiders has not gone away. In fact, since 1999—when the controversy was first ignited—it shows no signs of stopping at all. For example, on May 13, 2018, the *Mars Reconnaissance Orbiter* took a curious photo of a field of the spidery things. It was something that caught the attention of the world's media. CNN covered it extensively, as did the UK's *Daily Mail* newspaper, *Fox News*, and *Time*, among many others.

Although NASA dismisses the possibility that the Martian spiders and the banyan trees are evidence of incredible life on Mars, it's ironic that, by their own admission, their staff are looking deeply at how one day we, ourselves, may be able to alter the Martian soil for our very own benefit. For example, in 2005 NASA released the following on how it planned, one day, to hopefully be able to alter the Martian landscape, even to the point of growing plants on the surface of Mars:

> Take the cold tolerance of bacteria that thrive in arctic ice, add the ultraviolet resistance of tomato plants growing high in the Andes Mountains, and combine with an ordinary plant. What do you get? A tough plant 'pioneer' that can grow in Martian soil. Like customizing a car, NASA-funded scientists are designing plants that can survive the harsh conditions on Mars. These plants could provide oxygen, fresh food, and even medicine to astronauts while living off their waste. They would also improve morale as a lush, green connection to Earth in a barren and alien world.
>
> The plants would probably be housed in a greenhouse on a Martian base, because no known forms of life can survive direct exposure to the Martian surface, with its extremely cold, thin air and sterilizing radiation. Even then, conditions in a Martian greenhouse would be beyond what ordinary plants could stand. During the day, the plants would have to endure high levels of solar ultraviolet radiation, because the thin Martian atmosphere has no ozone to block it like the Earth's atmosphere does. At night, temperatures would drop well below freezing. Also, the Martian soil is poor in the mineral nutrients necessary for plants to thrive.[12]

Moving on to 2017, Gary Jordan of the NASA Johnson Space Center said:

In *The Martian* [a 2015 movie starring Matt Damon] Mark Watney uses the Martian soil to grow potatoes in the controlled environment of the "Hab." In reality, the soil on Mars actually *does* have the nutrients plants would need to survive on Mars! There may not be the right amount of nutrients depending on where astronauts land on the Red Planet, so fertilizers may need to be added to the soil. The perchlorates in the soil would be leached out and separated from the water. NASA is developing a simulant, a replication of Mars soil, to better understand how it can be used for plant growth and other purposes.[13]

In light of NASA's enthusiastic words, one day we may well be able to terraform Mars and turn it into a world thriving with life. If it isn't already. If we do, we'll likely find ways to introduce plants and trees into the environment by radically altering Mars's atmosphere. We'll melt the planet's polar icecaps to a non-catastrophic degree, significantly increasing the levels of CO_2 in the atmosphere and, in the process, heating up the planet, and, finally, making it habitable for us—possibly after we ravage our home world. After all, we're doing a very good job of wrecking the Earth, so we'll need somewhere else to live. It would be truly ironic, though, if the Martians had *already* attempted something near-identical when their world and society began to crumble, possibly millions of years ago. Just maybe, the banyan trees, the black spiders, and the Dalmatian spots that provoke so much controversy today are actually the last pieces of evidence of that ultimately doomed attempt by the Martians to save their world.

CHAPTER 1 2

"We Should Visit the Moons of Mars. There's a Monolith There."

||||||||||||||||||||||||||||||||||||||||||

arlier in this book, we addressed the subject of the twin moons of Mars, Phobos and Deimos. We also addressed how, in seemingly inexplicable, head-scratching fashion, author Jonathan Swift appeared to know of the existence of the two moons no less than a century and a half before they were officially discovered in the nineteenth century. The mystery doesn't end there, however. Indeed, the Martian moons are shrouded in just about as much mystery as the Red Planet itself, as we shall soon see. It was on the night of August 12, 1877, that astronomer Asaph Hall made history. That was the date on which it was confirmed that Mars had two small moons orbiting the planet. The discovery went down at the Naval Observatory in Washington, DC. As to their names, they come from Greek legend. Deimos was the much-feared deity of terror, while Phobos was the offspring of the Greek god of war, Ares, and Aphrodite. Neither moon can be considered large—quite the opposite, in fact. Phobos has a radius of approximately seven miles, while Deimos's radius doesn't even reach four miles. The moons may well have originally been positioned in the Asteroid Belt but might have been "captured" in the distant past by the mighty, gravitational pull of Mars itself. This theory is disputed, however, by the European Space Agency (ESA), who have said:

> One reason to suspect that Phobos is not a captured asteroid is its density. Analysis of Mars Express radio science data gave new information about the mass of Phobos based on the gravitational attraction it exerts on the spacecraft. The team concluded that Phobos is likely to contain large voids, which makes it less likely to be a captured asteroid. Its composition and structural strength seem to be inconsistent with the capture scenario.[1]

That the moons are filled with anomalies is not in dispute. Even NASA admits that much. For example, the surface of Phobos is covered in what NASA terms as *grooves*. Mac Tonnies chose to refer to them, in eye-opening fashion, as *roads*.[2] That is a far more provocative and image-creating word to use. It's also one that is all but guaranteed to provoke imagery of massive construction programs on Phobos millions of years ago. As you'll soon see, Tonnies suspected that such construction may well have gone on when the Martian civilization was still a thriving one.

Of those mysterious grooves, NASA says this:

> The long, shallow grooves lining the surface of Phobos are likely early signs of the structural failure that will ultimately destroy this moon of Mars. Orbiting a mere 3,700 miles (6,000 kilometers) above the surface of Mars, Phobos is closer to its planet than any other moon in the solar system. Mars' gravity is drawing in Phobos, the larger of its two moons, by about 6.6 feet (2 meters) every hundred years. Scientists expect the moon to be pulled apart in 30 to

50 million years. . . . Phobos' grooves were long thought to be fractures caused by the impact that formed Stickney crater. That collision was so powerful, it came close to shattering Phobos. However, scientists eventually determined that the grooves don't radiate outward from the crater itself but from a focal point nearby. . . . More recently, researchers have proposed that the grooves may instead be produced by many smaller impacts of material ejected from Mars. But new modeling by Hurford [Terry Hurford of NASA's Goddard Space Flight Center in Greenbelt, Maryland] and colleagues supports the view that the grooves are more like "stretch marks" that occur when Phobos gets deformed by tidal forces.[3]

It should be noted that there is no definitive answer to the "roads." In November 2018, Mike Wall, writing at Space.com, said that: "The weird linear grooves scoring the surface of the Mars moon Phobos were likely carved by boulders knocked loose by a giant impact, a new study suggests. That impact created Phobos' most notable feature—the 5.6-mile-wide (9 kilometers) Stickney Crater, which is about one-third as wide as the moon itself."[4]

Mac Tonnies told me: "We might be seeing [on Phobos] what's left of a huge excavation designed to propel mass into outer space; Stickney Crater might be the evidence of that. I'll get flack for saying this, but I don't rule out that Phobos itself is an alien spaceship—a natural moon engineered into something very different. Mega-scale reconstruction."[5]

The European Space Agency—the brains behind the ambitious *Mars Express* spacecraft program designed to explore Mars and its moons—notes that Phobos has a decidedly weird orbit:

Phobos has an equatorial orbit, which is almost circular. It orbits once every 7 hours 39 minutes just 5989 km above the surface of Mars. Its orbit is decaying by 1.8 cm per year, so it is expected to crash into Mars, or break up to leave a ring of fragments around the planet, within 100 million years. The orbital period of Phobos is three times faster than the rotation period of Mars, with the unusual result among natural satellites that Phobos rises in the west and sets in the east as seen from Mars. It orbits so close to the surface of Mars that the curvature of the planet would obscure its view from an observer standing in Mars' polar regions.[6]

One of those who recognized the weird orbit that Phobos displayed was a Russian astrophysicist named Iosif Shklovsky. Born in 1916 in the Ukraine, Shklovsky came to the startling conclusion that Phobos's odd orbit was provoked by the possibility that the moon had an extremely low density—something that could, in theory, at least, be caused by Phobos being hollow. Not only that, Shklovsky suggested that the hollowing of the tiny moon was most likely not natural—that, as Mac Tonnies also suggested, incredible mega-scale reconstructing by ancient Martians may have been the answer to the riddle. If this all sounds just too incredible—the idea of creating a world within a world—it's worth noting the words of Fred Singer. In 1954, he was the recipient, from President Dwight D. Eisenhower, of a White House Special Commendation for his achievements in the field of astronomy. Singer said of Shklovsky's theory:

If the satellite is indeed spiraling inward as deduced from astronomical observation, then there is little alternative to the hypothesis that it is hollow and therefore Martian made. The big "if" lies in the astronomical observations; they may well be in error. Since they are based on several independent sets of measurements taken decades apart by different observers with different instruments, systematic errors may have influenced them.[7]

Although Singer certainly did not endorse the "Phobos is hollow" theory, his words made it clear that he was at least open-minded on the matter.

Now, we come to what is perhaps one of the most controversial of all Phobos's anomalies: that of its very own "monolith," as it has been termed.

In 1968, the acclaimed sci-fi epic *2001: A Space Odyssey*, the work of producer and director Stanley Kubrick, hit cinemas all around the world and proved to be a great success. In the opening segment of the movie we see a tribe of primitive, early "ape-men" fighting for survival in the harsh landscape of what is now the continent of Africa—though in the film it plays out millions of years ago. One day the tribe awakes to find that overnight a strange, black-colored monolith has appeared in their very midst. In no time at all, the intelligence levels of the creatures are increased by, it appears, the monolith itself. The implication is that something of an extraterrestrial nature has kick-started human civilization into what will eventually be high gear. As amazing and as unbelievable as it may sound, a monolith has been photographed on the surface of Phobos. A case of fact eerily mirroring fiction? Well, that very much depends on one's own opinion.

We have a man named Efrain Palermo to thank for bringing this incredible matter to light. As someone with a deep interest in Mars and its moons, he decided to make an extensive review of the many pictures that NASA had secured from its *Mars Global Surveyor* in 1998. Palermo was amazed by what he found. He said that, while studying one particular image from Phobos ". . . my eye caught something sticking up out of the surface." As to what that "something" was, Palermo added: "I downloaded it into Photoshop and zoomed into that area, and there it was, an apparent cylindrical shaped object casting a longish shadow and having a slanted roof."[8]

Dr. Mark Carlotto, author of *The Martian Enigmas* and someone who has undertaken groundbreaking research into the controversies surrounding the Face on Mars and the Cydonia region, took an interest in the apparent monolith. Palermo explains that Carlotto ". . . referred me to Lan Fleming, a NASA imaging specialist who has interest in Mars and other solar system anomalies. Lan looked at it and upon further examination and study concluded as did I, that this was a physical anomaly on the surface of Phobos."[9]

That's putting it mildly.

Fleming worked at the Johnson Space Center and took a deep interest in the monolith, even going so far as to suggest that the monolith was not a random piece of oddly formed rock. Closer and clearer imagery analysis demonstrated that the monolith actually had a somewhat-pyramid-like shape, and not that of a typical monolith. The point, however, was that the monolith still appeared

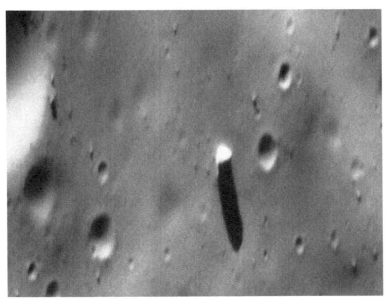

The oddest obelisk of all. (NASA)

to be the creation of some form of intelligence. Phobos lacks an atmosphere; thus there cannot be any kind of natural erosion on Phobos that might otherwise allow for the transformation of a regular rock into something radically different—and, something undeniably recognizable to us, too.

Fleming offered his views on the matter:

There seems to be no good reason to put the word "monolith" in quotes to describe this object. Efrain's interpretation is absolutely correct. At the very least, this object is a monolithic block of stone, although its high reflectivity may indicate that it is composed of something else. But how likely is it that it's artificial? That depends in part on how tall the object is relative to the width of its base. A block of stone several times longer than its height created from the impact of some large meteoroid would be unlikely to land on its narrow end and remain upright.[10]

One of those who took a big interest in the monolith on Phobos was the second man to walk on the surface of our Moon, Buzz Aldrin. He told C-SPAN: "We should visit the moons of Mars. There's a monolith there; a very unusual structure on this little potato-shaped object that goes around Mars once every seven hours. When people find out about that they are going to say, 'Who put that there? Who put that there?' Well, the universe put it there, or if you choose God put it there."[11]

Or, the Martians did.

CHAPTER 13

"A Beacon Erected by Aliens for Mysterious Reasons"

IIIIIIIIIIIIIIIIIIIIIIIIIIIIIIIIIIIIIIIIII

In 1998 the *Mars Global Surveyor* photographed what looks very much like a monolith on the surface of Mars's moon Phobos. And, as we have seen, there is a great deal of debate regarding what the photo shows—or, depending on one's perspective, what it does not show. The controversy surrounding the eye-opening image continues to annoy NASA and to amaze seekers of Martians and those who conclude that Mars was once a world filled with life. It should be noted, though, that Phobos's monolith is not alone. That's right: Phobos has a rival in the weird stakes. Say hello to the monolith of Mars. That's right: Both Phobos and its parent planet appear to have on their surfaces what seem to be obelisk-shaped stones. It's no surprise that the controversy surrounding the "object" seen on the Red Planet reverberated all around the world when the story reached the eager ears and eyes of the media.

Live Science said of this particularly interesting development in the matter of Martian mysteries:

> The object in question was first spotted several years ago after being photographed by the HiRISE camera onboard the *Mars Reconnaissance Orbiter*, a NASA space probe; every so often, it garners renewed interest on the Internet. But is it unnatural—a beacon erected by aliens for mysterious reasons, and even more mysteriously paralleled in the imaginations of Stanley Kubrick and Arthur C. Clarke, creators of "2001"? Or is this rock the work of nature?[1]

That was the question on the minds of just about every researcher of Martian anomalies.

Although it was in 2012 when the startling news of the amazing find took place, NASA has stated that the image was actually secured "several years" earlier. Regardless of where I, you, and the scientific community stand on the controversy, the fact is that the photo captured by NASA most assuredly *does* show something that looks like a monolith—hence the reason why it went on to quickly create such a controversy. And that is precisely also why it continues to do so. Not everyone is sure that the monolith is a monolith, after all, though. One of those who takes a down-to-earth explanation on all of this is Jonathon Hill. In his position at the Mars Space Flight Facility at Arizona State University, Hill suggested that what people were really seeing was "a roughly rectangular boulder." And a natural boulder, not something that was carved by intelligent beings millions of years ago. Hill explained why, in his opinion, there was very little—in fact, absolutely nothing—to get excited about: "When your resolution is too low to fully resolve an object, it tends to look rectangular because the pixels in the image are squares. Any curve will look like a series of straight lines if you reduce your resolution enough."[2]

Adding to the words of Jonathan Hill was the opinion of Yisrael Spinoza, a spokesperson for the HiRISE department of the University of Arizona's Lunar and Planetary Laboratory. Spinoza said: "It would be unwise to refer to it as a 'monolith' or 'structure' because that implies something artificial, like it was put there by someone, for example. In reality it's more likely that this boulder has been created by breaking away from the bedrock to create a rectangular-shaped feature."[3]

Alfred McEwen, also of the university, was not buying into the man-made (or, rather, Martian-made) explanation for what was being seen, either: "Layering from rock deposition combined with tectonic fractures creates right-angle planes of weakness such that rectangular blocks tend to weather out and separate from the bedrock."[4]

As for the work of HiRISE, NASA notes the following:

> The High Resolution Imaging [Science] Experiment is known as HiRISE. The big and powerful HiRISE camera takes pictures that cover vast areas of Martian terrain while being able to see features as small as a kitchen table. . . . HiRISE (High Resolution Imaging Science Experiment) has photographed hundreds of targeted swaths of Mars' surface in unprecedented detail. The camera operates in visible wavelengths, the same as human eyes, but with a telescopic lens that produces images at resolutions never before seen in planetary exploration missions. These high-resolution images enable scientists to distinguish 1-meter-size (about 3-foot-size) objects on Mars and to study the morphology (surface structure) in a much more comprehensive manner than ever before.[5]

Does it really need to be stated that not everyone bought into what, for many, was a perfectly acceptable explanation for the presence of the monolith-like piece of rock? Probably not, but let's see what those on the other side of the fence had to say about this distinctly odd situation. The *Daily Mail* newspaper chose to hark back in time. In doing so, they reminded people of what Buzz Aldrin, the second man to walk on the surface of the Moon, had said about the Phobos monolith: "We should visit the moons of Mars. There's

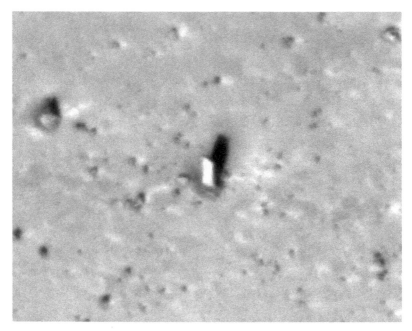
Monolith mysteries. (NASA)

a monolith there."[6] The *Daily Mail's* article was written in a fashion that clearly led the reader to conclude that whatever had been captured on film on Phobos was not alone. The surface of Mars itself, it now seemed, was echoing the secrets of Phobos.

So, where does all of this leave us? The reality of the situation is that it leaves us in what is very much a state of limbo. The staff of HiRISE have provided us with what is a perfectly plausible explanation for why it appears that there is a monolith-like protrusion on Mars. On the other hand, that there is a distinctly similar, obelisk-like object on the surface of one of Mars's moons, too, is something that we should not forget. In the same way that the Face on Mars, the D&M Pyramid, and the "Sphinx"-like block of stone captured on film in July 1997 near-hypnotized people with their strangely familiar—yet definitely out of place—appearances, it's difficult to fully dismiss the possibility that what we are seeing are some of the very last remnants of a nonhuman race of beings that died out when we were still small, hairy animals roaming the plains of what is now the continent of Africa.

Perhaps, one day, we will be in a position to embrace the theory that the many and varied structures on Mars, and on one of its two satellites, were not the work of nature, but the work of Martians.

CHAPTER 14

"Has Stonehenge Been Found on Mars?"

|||||||||||||||||||||||||||||||||||||||

Pareidolia—the brain's tendency to process random, chaotic imagery into something much more—has played an undeniably significant "role" in the quest to try and understand the nature and origins of at least *some* of the more unusual objects, artifacts, and structures that can be seen on the surface of Mars. There's no doubt that the Face on Mars, the D&M Pyramid, and various other anomalies in the Cydonia region *can* be relegated to the domain of pareidolia. And most certainly *have* been. There are also those curious, huge tubes that pepper the landscape, and Phobos's monolith, all of which have researchers at odds with each other. The most important question in this particular aspect of the quest to find the truth of the Martians, however, is this: *Should* they be written off as nothing more than tricks of the eye? The above-phenomena aside, there are four particularly intriguing "things" on Mars that for some researchers of the Mars mysteries are evidence that the planet was once the home of a thriving civilization of humanoids. One of them is a Stonehenge-like formation that we will address in good time. Two are pyramid-like findings. The other is what looks

The secrets of the Sphinx. (Wikimedia Commons)

Who built the Sphinx and when? (Wikimedia Commons)

incredibly like the Sphinx of Egypt. You thought it couldn't get any stranger? It just did exactly that.

Before we get deeply into the matter of *their* Sphinx—if that is what it is—it's most important that we first take a careful and close look at *our* Sphinx. All is most assuredly not what it appears to be. Why so? Because deeply addressing the matter of the legendary carving at Giza, Egypt, provides us with a plethora of astonishing parallels between the two—something that it's vital to understand when it comes to the matter of comparisons between what we see on Earth and that which we have found on the ravaged surface of Mars. First and foremost it must be noted that our Sphinx is steeped in mystery and intrigue; it's not just the strange formation on the Red Planet that provokes such major, thought-provoking controversy. With that said, let's now take a trip to Egypt.

The Sphinx stands out prominently against the Egyptian land-scape and the massive Pyramids that provoke so much wonder and mystery in those who visit it. The first thing to note are the Sphinx's dimensions. The Sphinx is close to 240 feet in length and is about 62 feet at its widest. As for its height, it's 66 feet. That's the easy part of the story. Much harder to determine is who built the Sphinx. In that sense, we need to focus on the who, the when, and the why of it all. In doing so we open the proverbial hornet's nest.

Speak to mainstream archaeologists and historians and they'll tell you that the Sphinx was constructed at some point between 2575 and 2465 BC. This was the period when the pharaoh Khafre ruled over the land. He was the fourth king of the fourth dynasty, someone who oversaw the construction of the second of the three legendary Giza Pyramids. Egyptologists suggest that the Sphinx dis-plays the face of Khafre. Maybe so. It should be noted, however, that matters are not quite so clean and cut as might be assumed, and as will soon become acutely obvious.

A detailed examination of the history of the Sphinx and its con-nections to Khafre strongly suggests that the amazing structure was not the work of the legendary pharaoh, after all. Rather, Khafre may very well have ordered the careful *modifying* of what was actually *a pre-existing* carving, one that may have been created by a race of people whose origins were completely unknown to Khafre and to the people of the fourth dynasty. And to us, too. In that sense, Khafre had in his hands an already-created Sphinx, one that only needed a relatively small degree of alteration to satisfy his largely ego-driven needs and demands. It should be noted that this is not at all a new concept. It has been flying around for years—albeit largely ignored by mainstream archaeologists, however. For example, such an admittedly controversial picture was in place way back in the 1940s. And it came from an archaeologist named Selim Hassan. The American University in Cairo says that Hassan was

one of the foremost Egyptologists of his time. After pursuing archaeological studies at the Royal College of Education, Hassan became one of the first Egyptians to secure a position as a curator at the Egyptian Museum in Cairo. He later received degrees in ancient languages and religions from the Louvre and Sorbonne in Paris, and a doctorate from Vienna University. The first Egyptian to hold Cairo University's chair in Egyptology, Hassan was a prolific scholar, publishing over 170 books and articles. During his long career he was involved in many excavations throughout Egypt, including the clearance of the Sphinx and its mastabas [a rectangular-shaped tomb] in Giza between 1929 and 1937.[1]

There's no doubt that the foremost expert in this field of the history of the Sphinx today is Robert M. Schoch, PhD. As his bio states, Schoch is

a full-time faculty member at the College of General Studies at Boston University and a recipient of its Peyton Richter Award for interdisciplinary teaching. Schoch earned his Ph.D. in Geology and Geophysics at Yale University in 1983. He also holds an M.S. and

M.Phil. in Geology and Geophysics from Yale, as well as degrees in Anthropology (B.A.) and Geology (B.S.) from George Washington University.[2]

In other words, Schoch sure knows his shit. And that includes the incredible things he has uncovered about the Sphinx and its history—which is at odds with the conventional theories.

In your author's own humble opinion, Schoch's most important discovery is that which concerns Egypt's rainfall. You might not think that rain could turn all that we know about the Sphinx on its head, but, amazingly, that is exactly what happened—if you buy into Schoch's conclusions, as I certainly do. Schoch has made a very good case that the Sphinx shows evidence of substantial weathering by rainwater. And we are not talking about the occasional drizzle here and there, but veritable torrents and for incredibly long periods of time. This causes significant problems for those who hang onto the conventional theory that the Sphinx was the creation of the workers that followed the orders of Khafre.

In Khafre's era, heavy rainfalls were unheard of. That situation is pretty much about the same today. The only possible explanation that makes any real sense is that the weathering by rainfall on the Sphinx occurred thousands of years before the fourth dynasty, when Egypt may have had substantial rainfalls. But, how many thousands of years? Schoch offers us something incredible: "Seismic data demonstrating the depth of weathering below the floor of the Sphinx Enclosure, based on my analyses (calibrated very conservatively), gives a minimum age of at least 7,000 years ago for the core body of the Sphinx (and more realistically, *on the order of 12,000 years ago* [author's emphasis])."[3]

As Schoch rightly notes: "Standing water in the Sphinx Enclosure would not accelerate the depth of weathering below the floor of the enclosure."[4] Is such a theory embraced by mainstream historians and archaeologists of ancient Egypt in general, and of the

Sphinx in particular? No, not at all. Those same historians and archaeologists, however, have yet to fly the flag of another, rational theory that would explain the evidence for massive rainfalls in ancient Egypt—and on the Sphinx.

There is yet another reason to believe that the Sphinx was created in the very *distant* past, rather than in the relatively *recent* past—at least, in terms of the history of the emergence of human civilization. It's a reason that is very difficult to dismiss: Many of the artifacts that we absolutely know were constructed in the era in which Khafre ruled, display *zero* evidence of having been weathered by near-endless rain. Why should the Sphinx display such undeniable weathering, when other fourth-dynasty items do not? The answer is clear: The Sphinx was built in an earlier era—in fact, an *incredibly* earlier era, as Schoch posits. Now, let's take a look at the matter of the Sphinx having been altered, probably under the orders of Khafre.

Is it possible that the Sphinx we see today is nothing less than a radically altered creation of some presently unknown civilization? Yes, it is possible; it's likely, in fact. One only has to take a brief look at the Sphinx to see that the head—which certainly does seem to resemble Khafre—is much smaller than it should be, when compared to the dimensions of the lion-like body of the iconic creation. On that latter matter, Robert Schoch has suggested that the original carving—mutilated and shrunken in the process to the orders of Khafre—may have been that of a lion, too. Moving back to the matter of rain, you should be aware that while the body of the Sphinx is undoubtedly water-weathered, the head is definitely not. How can that be? The answer is as obvious as it can possibly be: When Khafre chose to radically turn the head into a resemblance of himself, the old weathering that was done to the head was removed. The outcome was a smooth, unblemished visage—something akin to a ravaged, wrinkled, over-the-hill Hollywood star given a new lease on life and a smooth, new appearance thanks to Botox.

From Egypt to Mars and back again. (Wikimedia Commons)

Then, there's the equally mind-blowing matter of Nefertiti, of whom the *Ancient History Encyclopedia* notes:

> was the wife of the pharaoh Akhenaten of the 18th Dynasty of Egypt. Her name means, "the beautiful one has come" and, because of the world-famous bust created by the sculptor Thutmose (discovered in 1912 CE), she is the most recognizable queen of ancient Egypt. She grew up in the royal palace at Thebes, probably the daughter of the vizier to Amenhotep III, a man named Ay, and was engaged to his son, Amenhotep IV, around the age of eleven. There is evidence to suggest that she was an adherent of the cult of Aten, a sun deity, at an early age and that she may have influenced Amenhotep IV's later decision to abandon the worship of the gods of Egypt in favor of a monotheism centered on Aten. After he changed his name to Akhenaten and assumed the throne

of Egypt, Nefertiti ruled with him until his death after which she disappears from the historical record.[5]

As incredible as it may very well sound, there is an image on Mars that shows something that eerily mirrors a side image of Nefertiti, and that comes courtesy of the Jet Propulsion Laboratory, no less. On the Mars Artifacts site we learn the following of this curious development, which is practically jaw-dropping:

At Phoenicis Lacus, dark areas on the surface of the planet form a convincing face in profile with eye, nose, nostril, lips, chin, neck, and a very large hat with a headband. The eye is especially com-

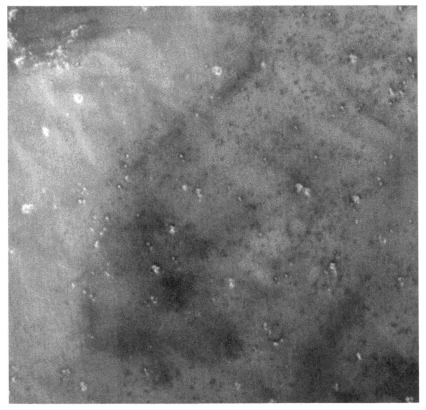

Nefertiti on Mars? (NASA)

pelling in that all the important parts of a human eye are present: a pointed lid aperture, light colored sclera (white of the eye), cheekbone, round iris, and pupil. Lashes appear to be laden as if with makeup. One cannot help but be reminded of Egyptian art because of the long Nefertiti-like hat. Similarities with the unfinished plaster bust of Nefertiti from Amarna, now in Berlin, are obvious.[6]

Even Mac Tonnies, who was noted for his ever-restrained approach to the matter of Mars's mysteries, was particularly impressed by the Nefertiti-like imagery that was released by the Jet Propulsion Laboratory. He said of the "strikingly Egyptian-looking 'bust'" that it "appears to consist of dark material on the Martian surface that's been deposited in a striking feminine likeness."[7]

Now, it's time for us to take a look at the matter of a connection—possibly even a *direct* connection—between the Sphinx at Giza, Egypt, and what may very well be the remains of an ancient, ravaged Sphinx on an equally ruined Mars. It's a controversy that taunts, tantalizes, and teases us with its incredible implication: that somehow, two worlds, millions of miles apart from each other, are intertwined to an incredible and history-challenging degree.

We have to thank the staff of NASA's Jet Propulsion Laboratory (JPL) for bringing a certain photograph to our attention—a photograph that is undeniably eye-catching. It seems to show a Sphinx and two ravaged pyramids. They have become known as the Twin Peaks. NASA could have quite easily obliterated the "offending" image, and no one outside of the space agency would have ever been any the wiser. The fact is, however, that NASA did not delete, shred, or burn the controversial image. What JPL personnel *did* do was to place the image firmly in the public domain. When questioned, they were wholly content to go with the pareidolia theory in relation to the Martian Sphinx. Publicly, it was no big deal to NASA. And it remains that way to them this very day.

Conspiracy theorists, however, might suggest another scenario: that NASA is engaged in a slow and deliberate program to, bit by bit, acclimate us to the idea that Mars was once a world not unlike ours—possibly even a world with inhabitants not too dissimilar to us. Such a theory, however, requires significant evidence to support it. So far, there is none. There's just a picture. We might say, though: But *what* a picture!

NASA and the Twin Peaks. (NASA)

For those who accuse NASA of "hiding the truth," it should be noted that it was the space agency itself that set the scene for allowing us to get on the trail of the Martian Sphinx controversy. In the JPL's very own words:

> The "Twin Peaks" are modest-size hills to the southwest of the *Mars Pathfinder* landing site. They were discovered on the first panoramas taken by the IMP camera on the 4th of July, 1997, and subsequently identified in Viking Orbiter images taken over 20 years ago. The peaks are approximately 30–35 meters (~100 feet) tall. North Twin is approximately 860 meters (2800 feet) from the lander, and South Twin is about a kilometer away (3300 feet). The scene includes bouldery ridges and swales or "hummocks" of flood debris that range from a few tens of meters away from the lander to the distance of the South Twin Peak.[8]

All that's missing is a Martian pharaoh. (NASA)

While NASA takes a "much ado about nothing" approach to what the image shows, there's no doubt that an issue sorely needs resolving, as we'll soon see. When I interviewed Mac Tonnies on the matter of the Martian Sphinx and the Twin Peaks, he told me the following:

> The region is not without its mysteries. In the *Pathfinder*'s landing charts is a third peak simply labeled "North Knob." And while it is certainly knob-like in shape, its perimeter is inscribed by a perfect square. The effect is obvious and uncanny and doesn't appear around the smaller Twin Peaks. North Knob, like the Twin Peaks, is situated on an extinct floodplain, so its square perimeter is doubly surprising; flood features on Mars are typical elliptical or teardrop-shaped, and one can easily tell the direction of the vanished flow by their alignment. The square perimeter may be due to a tough subterranean foundation; the irregular mass on top may be the remains of a structure of some kind that collapsed in the flood, scattering pieces of itself across the plain.[9]

Mac Tonnies was not the only one who recognized that there appeared to be a number of genuine mysteries contained in the *Pathfinder* photo: Mike Bara, a long-time researcher of the Martian enigmas and the author of *Ancient Aliens on Mars* and *Ancient Aliens on Mars II,* felt likewise. In a very well-argued online article titled "Mars Pathfinder Landing Site . . . A Sphinx Revisited?" Bara wrote:

> Besides all the "near field" *Pathfinder* anomalies, [geologist Ron] Nicks and [Richard] Hoagland began studying "super resolution" images of the two distant (almost one kilometer) "mountains" imaged on the horizon of the Landing Site: the celebrated "Twin Peaks." Nicks, in particular, soon realized that these features showed definite signs of engineering, as opposed to natural erosional processes. And, although he recognized that the area (as NASA had previously advertised) obviously had been subjected to "a brief but catastrophic flood," he could not explain some of the strikingly geometric features he was seeing on the "Peaks" as "just geology." There were obvious, repeating block-like structures (particularly on South Peak—see below), and some very unusual orthogonal (right angle) three-dimensional layering on the exposed ("downstream") surfaces.[10]

Now, we come to the really important part of the story. Mars researchers Mike Bara and Richard Hoagland say: "In the several years following *Pathfinder,* NASA scientists had been able to take multiple, overlapping images of the Pathfinder landing site and enhance them (via technique called Bayesian interpolation) well beyond their original resolution."[11]

In relation to the Sphinx and the enhancement of the imagery that created such a fuss, Bara said: "What had been merely a huge, dark, somewhat blobby shape on the original images suddenly emerged as recognizable on the new, highly processed images. It quickly became apparent that something very interesting might have been imaged on the Martian landscape, between South Peak and *Pathfinder* itself. What it seemed to be—at least to my eyes—was *a sphinx.*"[12]

Looking at the images and the enhancements that Bara has diligently studied, it's very easy to make a solid case that what we see in the distance, and with the South Twin Peak in view, is something that is very much Sphinx-like. That the Sphinx appears to be in front of severely eroded and damaged pyramid-type structures, only adds to the mystery and the intrigue—as does the work of Hoagland and Nicks. Exhibiting restraint, but clearly fascinated by the developments, Mac Tonnies said to me: "While I can't agree with Bara's certainty that the Sphinx is what he claims it is, the Sphinx remains a vague yet tantalizing shard of the Mars enigma. That there is something near the base of the Twin Peaks is irrefutable. Hopefully, future images taken at improved resolution will show us exactly what it is."[13]

We're still waiting.

The final words on this issue go to Tonnies: "There's a superficial similarity between some of the alleged pyramids in the vicinity of the Face and the better-known ones here on Earth. This has become the stuff of endless arcane theorizing, and I agree with esoteric researchers that some sort of link between intelligence on Mars and Earth deserves to be taken seriously."[14]

Now, let's move away from the matter of Mars and Egypt and turn our attentions to a connection between the Red Planet and Stonehenge, surely the world's most famous ancient stone circle of all.

On September 24, 2015, the UK's *Express* newspaper ran an article titled "Has Stonehenge Been Found on Mars? Ancient 'Alien' Stone Circle Discovered on Red Planet." That's quite a headline, to say the very least. "Mr. Enigma," who highlighted the breaking story on his YouTube channel, said that what appeared to be in evidence on the surface of Mars was "a perfectly circular platform with a strange cluster of stones emerging from it."[15] He added:

I know the formation is not an exact match, nor am I saying it is, indeed, a Stonehenge set up. I am just saying there is something strange about this area and it looks very much like the mysterious ancient stone circle of Stonehenge. Could the builders of Stonehenge have visited Mars and did they build the same thing there? Or did we have visitors who taught us how to build these things and do the same for long-lost beings on Mars as well? Or is this just another face on Mars illusion?[16]

This was excellent fodder for the media—particularly so for the tabloids. And despite the undeniably sensational headline that the *Express* staff chose to use to catch the eyes of readers, there is a genuinely serious issue here to be addressed. The fact is that the photo—from NASA, like so many others—shows what appear to be a number of large stones on the Martian surface, surrounded by a circle of much smaller ones. And while that circle is certainly not perfect, it's difficult to rationalize how such a formation could have been created solely by the enterprising work of Mother Nature.

Determined to outdo the *Express* in the popularity stakes, the UK's *Daily Mail*, in a series of articles, began referring to the curious formation as "Marshenge." English UFO researcher Nigel Watson told the *Daily Mail*:

Pyramid structures have been regularly spotted on Mars that have been linked to the ancient pyramids of Earth, now with Marshenge we have a link with our earlier prehistoric structures. The scorn of the skeptics have failed to erode such theories and with the discovery of Marshenge, it builds an even stronger case to suggest astronauts in the past visited our Solar System and had an enduring impact on human history. At present it is a huge leap of the imagination to compare Marshenge with the likes of Stonehenge, for example, we do not know the scale of these objects or what they look like in any detail. Up close they could prove to be random stones cast up by marsquakes—the equivalent of earthquakes—in its distant past.[17]

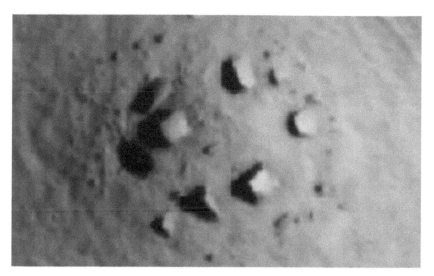

Stonehenge? No: Marshenge. (NASA)

It must be said that regardless of whether you embrace the
theory that there is a connection between Stonehenge and a
somewhat-similar stone formation on Mars, the fact is that the origins of Stonehenge are undeniably steeped in mystery—just like
the Sphinx of Egypt, let's not forget. Stonehenge, as it stands today,
is estimated to have been built around 3100 BC. That people lived
in the area as far back as 8000 BC, however, makes it very difficult
to determine the full picture and history of the famous creation.
The most baffling aspect of the overall controversy and mystery
surrounding Stonehenge revolves not so much around where the
massive stones stand, but where they came from. This is one of the
biggest enigmas surrounding Stonehenge.

Had the stones originated directly in what is now the English
county of Wiltshire—where the stones stand prominently to this
day—the construction of Stonehenge might have been a fairly easy
task to achieve. The fact is, though, that the huge bluestones, as
they are known, not only weigh in at about four tons each, but they
originated in the Preseli Mountains (Wales), which is where it's

believed the stones were quarried. The distance from those ancient mountains to the sacred site in Wiltshire that became known as Stonehenge is significantly more than 100 miles. Imagine trying to move such gigantic stones—multi-tons in weight and more than eighty in number—across mountains and rolling hills, and with nary a smooth, flat tarmacked road or a crane or several in sight. Such a thing would have been an incredible feat. One might be justified in saying it would be impossible. It would still be an incredible feat to this very day, never mind way back then. Why so? Because the bluestones were eclipsed by the much larger Sarsen stones, as they are known, *which range in weight from about 25 to approximately 50 tons.* Good luck making that trek! On top of that, there's the matter of then erecting the stones.

Notably, there is a strange and thought-provoking old tale that places the creation of Stonehenge into a very different context. Written in the 12th century by a character known as Geoffrey of Monmouth, *The History of the Kings of Britain* is a mighty tome that states Stonehenge was actually built by a mysterious race of giants who secured the stones from none other than the continent of Africa in times long, long gone. It must be said that the massive stones that were used in the creation of Stonehenge most assuredly did *not* originate anywhere in Africa. While that part of the story is undoubtedly false, it doesn't mean the whole tale lacks merit. According to Geoffrey, the giants had the ability to raise the mighty stones via what was called "The Giant's Dance." It was a ritual that involved the moving of the stones by the sound of music—a clear distortion of a concept known today as acoustic levitation, an utterly fascinating phenomenon we will return to later.

Were the giants of Stonehenge from the very same civilization of Goliath-like figures that were remote-viewed in 1984 by F. Holmes "Skip" Atwater, on behalf of the CIA? And the giant Martians imaged on the Moon by Ingo Swann in the 1970s—might that have been the reason as to why stone circles and Sphinxes can be found on both our home world and on Mars: that there is a connection between Mars and Stonehenge? Granted, it's an incredibly sensational theory, one many will dismiss outright and without a second thought. In finality on this specific aspect of the Martian mysteries, we should note that a great deal of the commentary surrounding the discovery of Marshenge was sensationalized. It should be noted, however, that the most important aspect of the story was not so much the tabloid-style sensationalism, but that the image came from NASA.

How many times do we have to be exposed to photos of non-terrestrial pyramids, sphinxes, and stone circles to make us realize that, in an era long gone, Mars and Earth shared something very special and something intimate—something that is now elusive, forgotten, and largely long gone? The answer is: when we have the undeniable smoking gun. Maybe we now have no less than four such smoking guns, in the form of Mars's Sphinx, the Twin Peaks, and Marshenge. Add to those all of the other many and varied oddities that have been photographed by NASA over the decades, and those four smoking guns may require being renamed as a full-blown arsenal.

CHAPTER 15

"A German Shepherd–Like Head"

||||||||||||||||||||||||||||||||||||||||

A s we have seen time and time again, there is strong evidence to support the theory that at least *some* of the anomalies found on the surface of Mars were the creations of Martians who are now extinct, who are living underground on their largely dead world, or who escaped the calamitous events that destroyed their civilization and made a new world for them on Earth in the distant past. The Cydonia region is a perfect example of this. It's comprised of the Face on Mars, the D&M Pyramid, and the "Fort," all of which provoke so much controversy and debate. And, let's not forget the saga of Mars's very own Sphinx and its close proximity to the "Twin Peaks" pyramids. All of the above offer us a great deal of food for thought. There are, however, other, less-impressive anomalies on the surface of Mars that have been offered as evidence of ancient life on Mars—and of intelligent life, too. That evidence is not, however, quite as persuasive as some might say, or who might want it to be. On the other hand, it's a fact that the weirder the Martian anomalies are, the more and more interest, publicity, and coverage they seem to get. *The X-Files'* "I want to believe" factor hard at work? To a significant degree, yes, probably.

With that said, let us now take a look at the data that has been put forward as evidence of life on Mars, but which we should be *extremely* careful of endorsing as the real deal. In this case, the data takes us from a NASA photo to the surface of Mars, and from Egypt to one of the Egyptians' most legendary gods. Before we get to the cases at issue, let's first take a greater look at what is the likely explanation: that of Pareidolia.

Larry Sessions, writing for *EarthSky*, gets right to the point of what Pareidolia really is:

> Maybe you've seen the proverbial bunny in a patch of clouds, or a clown's face in a mud splatter on the side of your car? Seeing recognizable objects or patterns in otherwise random or unrelated objects or patterns is called pareidolia. It's a form of apophenia, which is a more general term for the human tendency to seek patterns in random information. Everyone experiences it from time to time. Seeing the famous man in the moon is a classic example from astronomy. The ability to experience pareidolia is more developed in some people and less in others.[1]

Katherine Armstrong, in an article titled "Pareidolia," said:

> It is often hypothesized that people who are more religious, or believe in the supernatural, are more prone to pareidolia. Studies show that neurotic people, and people in negative moods, are more likely to experience pareidolia. The reason for this seems to be that these people are on higher alert for danger, so are more likely to spot something that isn't there. Women seem to be more prone to seeing faces where there are none. This may be linked to the fact that they have a better ability to recognize emotions through deciphering facial expressions.[2]

Yes, one can make a case that a hand, an arm, a shoulder, and even a head can be seen in the photograph. Clearly, though, this is without any shadow of a doubt a classic example of pareidolia. Not everyone in the Martian anomaly field agrees with me, though.

On March 24, 2017, the Emoluments of Mars blog ran an article titled "Mike Bara Gets it Wrong as Usual."[3] The opening words were as follows: "Mike Bara, the world-renowned geneticist and physiologist, was invited once again to blather for three hours on Jimmy Church's podcast *Fade to Black* this week. 90% of it was discussion of a series of images of odd-looking artifacts on Mars. There was the pistol, the sarcophagus, the fossilized dinosaur, and my personal favorite the 20ft high cat playing air guitar."

An upright cat, standing precariously on a huge hill, on none other than Mars? That is what one particular and peculiar photograph, secured by the *Mars Reconnaissance Orbiter*, seems to show. The word *seems* is the most important one in all of this. Provided to Mike Bara by an investigator named Malcolm Scott, the picture shows what looks like a thin, humanoid-type figure—which is so skinny it appears to resemble a classic stickman. We can see two arms (one of them raised), two legs, possibly two feet, and a head. Admittedly, the head is cat-like, and it has a pair of eyes and pointed ears. And, the "creature" appears to be standing in front of what looks like a shadowy entrance to a cave. In one sense the photo is actually quite striking—and undeniably surreal, too.

I showed a copy of the photograph to a well-known figure in the "ancient aliens" world who, on seeing it, practically foamed at the mouth. I'll avoid mentioning his name, as he has since moved on and relegated it to the "pareidolia basket." Before doing so, however, he spent a couple of weeks working on a paper titled "Anubis on Mars?" The reason: The person in question felt that the "upright cat" looked far more like none other than Anubis. And who, precisely, was Anubis? Let us take a look.

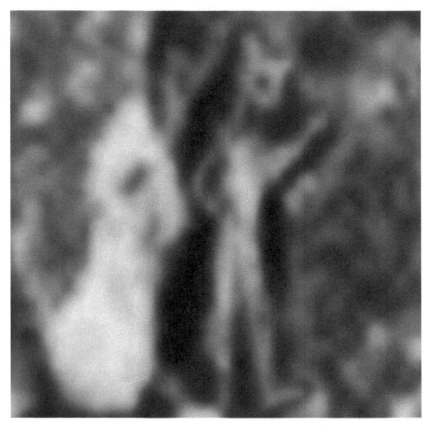

Anubis: A trick of the eye or something else? (NASA)

Crystalinks states:

Anubis is the Egyptian name for a jackal-headed god associated with mummification and the afterlife in Egyptian mythology. Anubis was the god to protect the dead and bring them to the afterlife. He was usually portrayed as a half human, half jackal, or in full jackal form wearing a ribbon and holding a flail in the crook of its arm. The jackal was strongly associated with cemeteries in ancient Egypt, since it was a scavenger which threatened to uncover human bodies and eat their flesh. The distinctive black color of Anubis "did not have to do with the jackal, per se, but with the color of rotting flesh and with the black soil of the Nile valley, symbolizing rebirth."[4]

Then, there are the words of Natalia Klimczak:

To date, archaeologists have not unearthed any monumental temple dedicated to this god. His "temples" are tombs and cemeteries. The major centers of his cult were located in Asyut (Lycopolis) and Hardai (Cynopolis). His name appears in the oldest known mastabas (mud-brick tombs) of the First Dynasty and several shrines to the god have been found. For example, a shrine and a cemetery of mummified dogs and jackals was discovered at Anubeion, a place located to the east of Saqqara. It seems that during the reign of the first dynasties he was even more significant than Osiris. This changed during the Middle Kingdom period, but Anubis continued to be one of the most important deities.

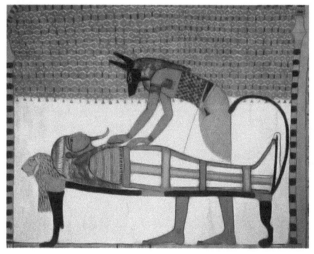

The Egyptian god of mummification. (Wikimedia Commons)

While most researchers who have bothered to look at the photograph have suggested that the "figure" in the photo is cat-like, my colleague was inclined to suggest that the "head" was somewhat jackal-like—which, to a degree, it is. Mike Bara said that the figure seemed to be holding what looked like . . . a guitar! That same colleague, however, got it into his mind that the "guitar" was actually an ancient weapon known as a flail. Anubis, it just so happens, was often portrayed in ancient artwork as "holding a flail in the crook of its arm."[5]

So, in this story we have a cat-like creature that might be holding a guitar, or Anubis holding what is perhaps a flail. Or, we just have random rocks that don't look quite so random, after all. We also have a connection to Egypt—a country that pops up over and over again when it comes to addressing the Martian anomalies, such as the Face on Mars.

The Anubis controversy continues. (Wikimedia Commons)

For a while, the man working on the paper at the time was quite excited by the whole story, not to mention the implications suggesting an Egypt-based link to Mars and Anubis. What was it that led him to drop his research on this particular matter? The cartoonish, skinny nature of the figure, that's what. The paper was finished, but never published, even though it does admittedly make for fascinating reading—if you're willing to stretch your imagination to a fairly significant degree, that is. So, whether you believe that Anubis-type creatures live on Mars or not, the fact is that this one, solitary photo provoked a well-known author and researcher to spend considerable time trying to make a case in favor of the photo. Such is the incredible, powerful lure that pareidolia has on people.

The story, however, is not quite over. One can make an argument that it has just begun. It revolves around a modern-day creature that is somewhat Anubis-like and that has become known as "the Dogman." And, there's an extraterrestrial aspect to the story, too. Just maybe, the story isn't quite all over. Yet.

Since 1991, the Wisconsin town of Elkhorn has been the lair and hunting ground of a terrifying creature that is the closest thing one can imagine to a real-life werewolf. The monster has become known as the Beast of the Bray Road, on account of the fact that many of the initial sightings were made on that particular road in the town. Without doubt, the expert on the monster is author and journalist Linda Godfrey. Some years ago, I interviewed Linda about her research into this malignant beast, which has become known as the Dogman. Since 1991, reports have surfaced from other parts of Wisconsin. Michigan, too. Linda told me[6]:

> The story first came to my attention in about 1991 from a woman who had heard rumors going around here in Elkhorn, and particularly in the high school, that people had been seeing something like a werewolf, a wolf-like creature, or a wolf-man. They didn't really know what it was. But some were saying it was a werewolf. And the werewolf tag has just gotten used because I think that people really didn't know what else to call it.
>
> I started checking it out. I talked to the editor of *The Week* newspaper here, and which I used to work for. He said, "Why don't you check around a little bit and see what you hear?" This was about the end of December. And being a weekly newspaper that I worked for, we weren't really hard news; we were much more feature oriented. So, I asked a friend who had a daughter in high school and

she said, "Oh yeah, that's what everybody's talking about." So, I started my investigations and got one name from the woman who told me about it. She was also a part-time bus driver.

In my first phone call to the bus driver, she told me that she had called the County Animal Control Officer. So, of course, when you're a reporter, anytime you have a chance to find anything official that's where you go. I went to see him and, sure enough, he had a folder in his file draw that he had actually marked *Werewolf*, in a tongue-in-cheek way.

People had been phoning in to him to say that they had seen something. They didn't know what it was. But from their descriptions, that's what he had put. So, of course, that made it a news story. When you have a public official, the County Animal Control Officer, who has a folder marked *Werewolf*, that's news. It was very unusual.

It just took off from there and I kept finding more and more witnesses. At first they all wanted to stay private, and I remember talking about it with the editor and we thought we would run the story because it would be over in a couple of weeks. The story was picked up by Associated Press. Once it hit AP, everything broke loose, and people were just going crazy. All the Milwaukee TV stations came out and did stories, dug until they found the witnesses, and got them to change their minds and go on camera, which some of them later regretted. And which I kind of regret, because it really made them reluctant, and kind of hampered the investigation.

They were all mostly saying that they had seen something which was much larger than normal, sometimes on two legs and sometimes on four, with a wolfish head. Some described it as a German Shepherd–like head, pointed ears, very long, coarse, shaggy, and wild-looking fur. One thing they all mentioned was that it would turn and look at them and gaze fearlessly or leer at them, and it was at that point that they all got really frightened. Everybody who has seen it—with the exception of one—has been extremely scared because it's so out of the ordinary. It was something they couldn't identify and didn't appear to be afraid of them. It would just casually turn around and disappear into the brush. It was never just out in the open where it didn't have some sort of hiding place. There was always a cornfield or some brush or some woods. So, that was pretty much the start of it.

Once that got out, I started finding other people who called me and got in touch with me and I sort of became the unofficial clearinghouse. And we called it the *Beast of Bray Road* because I've always been reluctant to call it a werewolf. The original sightings were in an area known as Bray Road, which is outside of Elkhorn.

Everybody seems to have an opinion about this that they are eager to make known and defend. I personally don't think there are enough facts for anybody to come to a conclusion. I have a couple of dozen sightings, at least. A few of them are second-hand and they date back to 1936. And they aren't all around Bray Road. Quite a number are in the next county, Jefferson. I've had a woman write me who insists it's a wolf. And I think a lot of people subscribe to that theory; yes, it's definitely a wolf and can't be anything else. But that doesn't explain the large size.

We've had all sorts of theories; mental patients escaping or some crazy guy running around. A hoaxer is another theory; that it's somebody running around in a werewolf suit. One or two could have been that, but I tend to have my doubts about that, because the incidents are isolated and not close together. One of the sightings was on Halloween, but that's also one of the people who got a really good look at it and they're sure it wasn't a human in a costume. Otherwise, most of them have been in really remote locations where, if you were going to hoax, the person would have to have been sitting out there in the cold just waiting for somebody to come along. So, if it's a hoaxer, my hat's off to them. But I tend not to think that's the case. I don't rule it out completely because once publicity gets out, things like that can happen.

Two hunters quite a bit farther north saw what looked like two "dog children" standing up in the woods. They were too scared to shoot when they saw them. They were not tall; they were juvenile-looking, standing upright, which is what scared them. But, otherwise, it's a single creature. Most of the sightings I receive aren't recent, and so people can't remember too well what the moon was like. But most of the sightings occur around the fall when the cornfields get big and there's really good hiding cover. So, that's anywhere from late August through November. And I've had some sightings from the spring. But there are other theories as well for what is going on.

Occasionally I'll get letters from people who say they are lycan-thropes themselves and their theory is that this is an immature, real werewolf and it cannot control its transformation, and that's why it allows itself to be seen occasionally. They are completely con-vinced of that. And there are people who believe it's a manifesta-tion of satanic forces, that it's a part of a demonic thing. They point to various occult activities around here. There are also people who try to link it to UFOs. Then there's the theory it's just a dog. One woman, a medium, thought that it was a natural animal but didn't know what it was. And there are a lot of people out here that do wolf-hybridizations, and I've thought to myself you'd get something like that. But that doesn't explain the upright posture. Then there's the theory that it's a creature known as the Windigo or Wendigo, which is featured in Indian legends and is supposedly a supernatural creature that lives on human flesh. But none of the descriptions from the Windigo legends describe a creature with canine features.

There's another possibility: I think a lot of these people are see-ing different things. And that when they heard somebody else talk about something, there's a tendency to say, "Oh, that must be what I saw." There's really no way to know. And there are differences in some of the sightings. I've had people ask me, "Are you sure this isn't Bigfoot?" Most of the sightings really don't sound like what people report as Bigfoot. But a couple of them do. There's one man who saw it in the 1960s in a different area of the county, who insists positively that he saw a Bigfoot, but doesn't want anyone saying he saw a werewolf. And the terrain around here isn't really the typical sort of Bigfoot terrain of forests where people usually report these things. We do have woods and a big state forest, but it's a narrow band of forest. It's a lot of prairie and is not what you would think a Bigfoot would live in. But you never know. I've also had the baboon theory, which I find extremely unlikely.

Of her first book on the subject, *The Beast of Bray Road*, Godfrey said to me: "Part of the angle of the book is looking at this as a sociological phenomenon and how something that a number of people see turns into a legend. And it has become that, a little bit. Personally, I'm still happy to leave it an open mystery. I don't have a feeling that it has to be pinned down."

A quarter of a century after her investigations into the mystery of the Beast of Bray Road began, Linda Godfrey's research and writing continue, and the sightings of the unearthly monster continue. Her published works now include *The Beast of Bray Road, Werewolves, Hunting the American Werewolf, The Michigan Dogman*, and *Real Wolfmen*. Collectively, they demonstrate something incredible: Werewolves may not be creatures of fantasy, legend, and mythology. They just might be all too real. And Bray Road may be just the place to see them.

■■■

While there are numerous theories for what the Dogmen are (demonic entities, shape-shifters, a rare/unique kind of wolf; the list goes on) the strangest of all the theories is that which suggests the creatures are extraterrestrials. Yes, you did read that right.

In 2005, Godfrey was contacted by a man—a military whistle-blower, we might say—who was an expert in the field of remote viewing. According to Godfrey's Edward Snowden–like source, the US government has uncovered data suggesting that the Dogmen are a very ancient, alien race that closely resembles the ancient deity of the Underworld. And who might that be? It's Anubis, that's who. Godfrey's informant also discovered—via remote viewing—that the Dogmen can "jump" from location to location via portals or doorways in the fabric of space and time. That's quite a story told to Godfrey: Dogmen from the stars that have a connection to Anubis—as well as a bizarre connection to Mars, all thanks to a NASA photo.

CHAPTER 16

"It May Be a Crab-Like Animal"

||||||||||||||||||||||||||||||||||||||

Now, let us take a careful look at some of the other imagery that has been collected, studied, and placed into the public domain as evidence that there was (and possibly still is) life on Mars. One particular case, which I personally think has a high degree of merit attached to it, concerns what has become known as the "Face-Hugger photograph." Taken in July 2015 by NASA's *Curiosity Rover*, it appears to show a strange creature that looks astonishingly like the monstrous face-invading creatures that appear in the phenomenally successful series of *Alien* movies. They starred Sigourney Weaver as Warrant Officer Ellen Ripley and reaped in an incredible amount of money. The story broke in early August 2015. The headlines were predictably sensational. The UK's *Metro* newspaper ran with the story and titled their feature as follows: "Crab-Like Alien 'Facehugger' Is Seen Crawling out of a Cave on Mars."[1]

The article included the words of Seth Shostak, a skeptic when it comes to the matter of life on Mars, and the senior astronomer and director of the Center for SETI Research. ("SETI" stands for "search for extraterrestrial intelligence.") Shostak said of the strange-looking thing and other allegedly anomalous photos that reach him from time to time: "Those that send them to me are generally quite excited, as they claim that these frequently resemble something you wouldn't expect to find on the rusty, dusty surface of the Red Planet. It's usually some sort of animal, but occasionally even weirder objects such as automobile parts. Maybe they think there are cars on Mars."[2]

On the other side of the coin were the words of Scott Waring who, at *UFO Sightings Daily*, said: "It does appear alive. It may be a crab-like animal, or it also may be a plant. This object has many arms and one of them goes to the left of the picture a very long ways. That arm is longer than all others. Plant or animal it really doesn't matter. The significance of this is that it shows signs that it is alive. That is everything, but not to NASA."[3]

A spidery, crab-like thing prowls the harsh Martian landscape. (NASA)

And, that's pretty much how the online debate went on: For the world of science the face-hugger was nothing but an example of pareidolia. UFO seekers and conspiracy theorists said otherwise, suggesting that what looked like an eerie, creepy animal was *exactly* that: an eerie, creepy animal. That it appeared to be in the opening to a small cave and maneuvering up (or, granted, scuttling down) the cave wall with its multiple limbs only seems to add to the theory that NASA had captured on camera a genuine Martian animal— and a very hair-raising one, too! In my view, the spidery, crab-like animal is *not* a case of the eye seeing what it sorely wants to see. And, the story isn't quite over: Later, we'll take a close and careful look at the incredible story of a skilled remote viewer who, in late 2019, was able to add further, and undeniably astonishing, data to this particular story of the closest things to real-life face-huggers.

Now, what about Mars's very own dolphin? Say what? Yes. Or not. This is one of the examples that I find difficult to impossible to

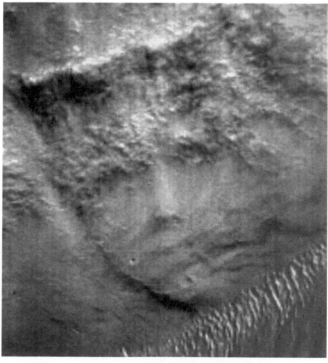

A king and his crown. (NASA)

believe is anything but a mark on the landscape that just happens to *appear* dolphin-like—which, admittedly, it does. That it is located not far from the Face on Mars has ensured that it still provokes debate and commentary. But, it really shouldn't. A carved dolphin? Not a chance in Hell. Forget it. Moving on, there is what has been termed "The Crowned Face." A photo taken in the Libya Montes area of a ring of mountains on Mars by the *Mars Global Surveyor* appears to show a large face, with a pointed chin, a pair of eyes, and a nose—and also, what appears to be crown-like headgear. Hence the name that has been attached to it, whatever "it" may be.

On this one, I can go both ways: I'm not convinced that it shows evidence of huge, mega-sculpting of the landscape. On the other hand, I definitely think it is well worth tackling this enigma to a greater degree, if only to relegate it to the "maybe" box. I should stress, though, that Mars expert Tom Van Flandern was quite enthusiastic about the discovery: "While not near the Cydonia area, this face portrayal is again striking for the richness of its detail, far better than the typical face arising in clouds or geological formations on Earth. The latter tend to be distorted and grotesque when they are more than simply impressionistic."[4]

"Dinosaur skulls" on Mars don't impress me in the slightest. I have yet to see one that really jumps out at me. There are a few of them, again captured by NASA's cameras, but I am not persuaded here. That said, there is one skull-like item that I find very interesting. Obtained by NASA's *Spirit*, the photo does appear to show a large skull on the rocky, desert floor of Mars. One can see a pair of eye sockets. Moreover, those same sockets seem to be in perfect alignment with each other. A nose and nostrils, a robust jaw, and even the vague outline of a bony mouth are all in evidence. The skull is clearly somewhat different to a regular, human skull, in the sense that it seems to be much bulkier. If the anomaly is what it appears to be—the skull of a Martian—it's hardly likely to look *exactly* like the skull of a member of the human race! There is something else too:

what look like a pair of protrusions—or what Mike Bara refers to as "jowl-like appendages"—situated around the chin area.[5] They look as if they could inflict serious injury if one got too close.

Now we come to a story that practically circled the globe in 2008. It provoked sensational comments, controversial observations, and wild conclusions. The story suggested that nothing less than the ancient statue of a man wearing a robe, or a woman wearing a long dress—with one arm poking out—could be seen on the Martian landscape. Also, the statue appeared to be in a sitting position, taking a break from whatever it is that the average Martian does on the average Martian day. It was a story that was picked up by not just the blogs of UFO researchers, but by the likes of CNN, the BBC, and Reuters. And the initial data seemed to be impressive. It would not stay like that.

Benjamin Radford, of the *Skeptical Inquirer,* and someone who has yet to see an anomaly that he cannot explain, said: "Accord-

A controversy-filled photo excites the media. (NASA)

ing to astronomer Phil Plait of *Bad Astronomy* website, if the image really is a man on Mars, he's awfully small: 'Talk about a tempest in a teacup!' Plait said: 'The rock on Mars is actually just a few inches high and a few yards from the camera.'"[6]

Plait was correct: Despite the extensive coverage that the story got, very few media outlets pointed out that the figure was indeed barely a few inches tall and that the *Spirit*—which secured the image—was actually extremely close to the "figure" when the photo was taken. Just about all reproductions of the photo made it appear as if the viewer was looking at a large object, or a human-sized figure, in the distance.

Mike Bara had an answer for this. He theorized that the admittedly eerie-looking figure was an ancient, Martian artifact of small proportions. To bolster his argument, Bara showed examples of equally small, carved figures of the Egyptian pharaoh, Khufu.[7] A very small Martian humanoid taking a break on a rock? A tiny piece of rock that seemed to resemble a living being? A carefully fashioned, small statue of the type that Mike Bara offered as a potential answer to the controversy? The answers, however, depend very much upon your own personal perspective. And still on the matter of perspectives: If only the mainstream media had made it abundantly clear from the beginning that what we were seeing was nowhere near a life-sized figure—but something around the size of a kid's toy soldier—we would not still be debating this issue years later—and a significant portion of this chapter could have been seamlessly omitted.

CHAPTER 17

"Liquid Water Flows Intermittently on Present-Day Mars"

||||||||||||||||||||||||||||||||||||||||

If there is one thing that all of us require to live, it is water. Without it, we are utterly doomed—as are plants, trees, flowers, and every living creature on the planet. But, what, exactly is water and why is it so important to our very survival? If life does exist on Mars, and if it even exists in a form broadly similar to us, then it is all but certain that the Martians would need massive amounts of water to allow them to not just survive, but to thrive, too—and possibly for untold millennia. And if the Face on Mars is not a spectacular example of pareidolia, but, instead, is a representation of how the Martians look (or looked), then that's a further reason to conclude that the Martians needed water. After all, the Face is strikingly human-like. So, if the appearance of the Martians was broadly similar to ours, then logic would dictate they—like us—would require regular and substantial amounts of water per day. The problem, however, is that so far we see very little evidence of water on Mars. Yes, there are huge amounts of water on Mars in the form of ice. That, however, is very different from being able to provide regular, clean water supplies to potentially millions of Martians, whether in eras long gone or today. Let us take a look at what NASA has to say about the matter of Mars and water.

On September 28, 2015, NASA announced the following in a press release:

New findings from NASA's Mars Reconnaissance Orbiter (MRO) provide the strongest evidence yet that liquid water flows intermittently on present-day Mars. Using an imaging spectrometer on MRO, researchers detected signatures of hydrated minerals on slopes where mysterious streaks are seen on the Red Planet. These darkish streaks appear to ebb and flow over time. They darken and appear to flow down steep slopes during warm seasons, and then fade in cooler seasons. They appear in several locations on Mars

The search for water on Mars. (NASA)

when temperatures are above minus 10 degrees Fahrenheit (minus 23 Celsius), and disappear at colder times.

"Our quest on Mars has been to 'follow the water,' in our search for life in the universe, and now we have convincing science that validates what we've long suspected," said John Grunsfeld, astronaut and associate administrator of NASA's Science Mission Directorate in Washington. "This is a significant development, as it appears to confirm that water—albeit briny—is flowing today on the surface of Mars.

"These downhill flows, known as recurring slope lineae (RSL), often have been described as possibly related to liquid water. The new findings of hydrated salts on the slopes point to what that relationship may be to these dark features. The hydrated salts would lower the freezing point of a liquid brine, just as salt on roads here on Earth causes ice and snow to melt more rapidly. Scientists say it's likely a shallow subsurface flow, with enough water wicking to the surface to explain the darkening."[1]

Although this was big news when the story broke, the fact is that evidence for water on Mars is nothing new. For example, NASA's *Mariner 9* probe, which headed to Mars on May 30, 1971, uncovered highly persuasive evidence that Mars was not a dead, stony, dusty world. There was water on the Red Planet. And if there was water, then just maybe there was life on the planet, too. *Mariner 9* sent back to NASA quite literally thousands of photographs of the Martian surface. A number of those same pictures showed what were undeniably ancient river bends and—equally undeniably—evidence of the erosion of the landscape by water. Then, there is the matter of NASA's *Viking 1* and *Viking 2*, both of which were launched in 1975. The former was designed to capture numerous photos of Mars from an orbital position, while the latter landed on the Martian surface, secured close-up imagery, and studied the alien landscape. As for the photos that both unmanned probes secured, they too found persuasive data that Mars had once been a planet with copious amounts of water. The *Viking* crafts, too, demonstrated that what was almost certainly erosion by water had occurred in the past. How far in the past? That question remained unanswered.

When NASA's *Mars Global Surveyor* craft was launched in November 1996, there were high hopes—*very* high hopes—within NASA that the probe would be able to expand upon what was then known about the controversy surrounding Mars and its water supplies. To a sensational degree, success was achieved. Among its many tasks, the *Surveyor* sent back to Earth images of the Centauri Montes and Terra Sirenum. The former is a mountain range, and the latter is a large area in Mars's southern hemisphere. Incredibly, both showed evidence of what was once a presence of incredible amounts of water. The findings at Terra Sirenum were particularly spectacular,

Ice at Mars's north pole. (NASA)

with a consensus that a huge, now-dry lakebed in the area had once held levels of water to around 600 feet deep. Moving onto NASA's *Mars Pathfinder*, which touched down on Mars in 1997, there was further data suggesting that Mars hadn't always been "dead."

Certainly, one of the most amazing developments came in May 2002. It was all thanks to the team that designed, created, and dispatched to the Red Planet the *Mars Odyssey* craft. In a truly eye-opening development, NASA excitedly revealed something that could not fail to capture the attention of just about anyone and everyone with an interest in Mars and its mysteries. NASA said: "Using instruments on NASA's 2001 *Mars Odyssey* spacecraft, surprised scientists have found enormous quantities of buried treasure lying just under the surface of Mars—enough water ice to fill Lake Michigan twice over. And that may just be the tip of the iceberg."[2]

Jim Garvin, a Mars Program scientist at NASA Headquarters (Washington, DC), added his words on the exciting development:

We have suspected for some time that Mars once had large amounts of water near the surface. The big questions we are trying to answer are, "where did all that water go?" and "what are the implications for life?" Measuring and mapping the icy soils in the polar regions of Mars as the *Odyssey* team has done is an important piece of this puzzle, but we need to continue searching, perhaps much deeper underground, for what happened to the rest of the water we think Mars once had.[3]

In March 2004, NASA prepared a press release with the following, stunning title: "Volcanic Rock in Mars' Gusev Cater Hints at Past Water." NASA stated:

NASA's *Spirit* has found hints of a water history in a rock at Mars' Gusev Crater, but it is a very different type of rock than those in which NASA's Opportunity found clues to a wet past on the opposite side of the planet. A dark volcanic rock dubbed "Humphrey," about 60 centimeters (2 feet) tall, shows bright material in interior crevices and cracks that looks like minerals crystallized out of water, Dr. Ray Arvidson of Washington University, St. Louis, reported at a NASA news briefing today at NASA's Jet Propulsion Laboratory, Pasadena, Calif. He is the deputy principal investigator for the rovers' science instruments.[4]

As we have seen, and as NASA has demonstrated time and time again, the massive amounts of water that can be found on Mars in the form of ice and are incredible. And, almost certainly, Mars was home to massive amounts of running water in eras that have yet to be determined. There is also the matter of a very small presence of water vapor on Mars—or, what was *assumed* to have been small amounts of water vapor. As will now become clear, there was more water vapor than had been assumed. October 2011 was the time when a new angle to the "water on Mars" controversy hit the headlines. *Astrobiology Magazine* informed its readers of the following words:

New analysis of data sent back by the SPICAM spectrometer on board ESA's [European Space Agency] *Mars Express* spacecraft has revealed for the first time that the planet's atmosphere is supersaturated with water vapor. This surprising discovery has major implications for understanding the Martian water cycle and the historical evolution of the atmosphere. This information is valuable in determining whether or not Mars once supported habitable environments for life.[5]

Now, there's the most important aspect of this controversy: that of water in the form of liquid. For years—decades, in fact—it was assumed that Mars completely lacked water. That assumption, however, all changed in the summer of 2018. That was when it was revealed that a body of water had been found on the Red Planet, specifically below the south polar icecap, and at a depth of about one mile. To say that this was an unforeseen development and an amazing revelation is an understatement of spectacular proportions. On this occasion, the discovery was not made by NASA, but by Marsis, which is radar technology designed and utilized on the European Space Agency's *Mars Express* orbiter.

The driving force behind the project was Professor Roberto Orosei, of the Italian National Institute for Astrophysics. He stressed to the world's eager media that the lake was not overly large. It was also noted, however, that the team was not talking about a small pool of water, either. No, they were talking about nothing less than a lake that had an approximate width of twelve miles. And, at the very least, the lake was believed to have been about three feet deep. This begs an important question: Why hadn't such a large body of water been discovered earlier?

The answer is very simple, but perhaps unforeseen: The lake is not actually on the surface of Mars. Rather, it is positioned below the south polar ice cap. Not quite a case of hiding in plain sight, but you get the point. It took the Marsis technology to find the lake. As the BBC explained: "Radar instruments like Marsis examine the surface and immediate subsurface of the planet by sending out a signal and examining what is bounced back."[6] In this particular case, the science and technology behind Marsis revealed the discovery of the lake.

Clearly enthused by the discovery, Professor Orosei added: "This really qualifies this as a body of water. A lake, not some kind of meltwater filling some space between rock and ice, as happens in certain glaciers on Earth." The professor had more to say: "In light blue you can see where the reflections from the bottom are stronger than surface reflection. This is something that is to tell us the telltale sign of the presence of water."

The big question, of course—the one that just about everyone wanted answering—was this: If the lake was made up of a significant body of water, was it possible that it was the home to some form of life? Even the findings of microbes would amount to an incredible discovery. So, where did things stand at that point? The Open University's Dr. Manish Patel provided the answer to that particular question: "We have long since known that the surface of Mars is inhospitable to life as we know it, so the search for life on Mars is now in the subsurface. This is where we get sufficient protection from harmful radiation, and the pressure and temperature rise to more favorable levels. Most importantly, this allows liquid water, essential for life."

The BBC interviewed Dr. Patel, who said: "We are not closer to actually detecting life, but what this finding does is give us the location of where to look on Mars. It is like a treasure map—except in this case, there will be lots of 'X's marking the spots." In other words, life had not been found yet, far below the south polar icecap, but this was certainly a major leap forward in the quest to find life on Mars. There were, however, several downsides to all of this. The *Mars Express* team explained that the cold temperature might have an adverse effect on any kind of life that might exist in the lake. That the temperature in the lake was estimated to be around −10 to −30 degrees Celsius would likely provide problems for even the most primitive forms of extraterrestrial life.

On this matter, Dr. Claire Cousins, an astrobiologist from the UK's University of St. Andrews, said: "It's plausible that the water may be an extremely cold, concentrated brine, which would be pretty challenging for life." This was far from the end of the matter, however. Everyone involved in the project had high hopes that the research into the lake and its contents would continue at a steady pace.

The Open University's Dr. Matt Balme told the media: "What needs to be done now is for the measurements to be repeated elsewhere to look for similar signals, and, if possible, for all other explanations to be examined and—hopefully—ruled out. Maybe this could even be the trigger for an ambitious new Mars mission to drill into this buried water-pocket—like has been done for sub-glacial lakes in Antarctica on Earth."

Professor Orosei reached his conclusions: "Getting there and acquiring the final evidence that this is indeed a lake will not be an easy task. It will require flying a robot there which is capable of drilling through 1.5km of ice. This will certainly require some technological developments that at the moment are not available."

We have addressed the matter of huge amounts of running water in Mars's ancient past. We've also taken a close look at the matter of Mars's frozen water supplies today. And, let's not forget the "super-saturated" nature of the water vapor that is present on Mars. But, what about Mars's ocean? Yes, an ocean. *Of water.* March 5, 2015, was the date NASA brought that particularly controversial matter to the attention of the world. NASA is not known for being an agency that speculates to any significant degree—quite the opposite, in fact. On this occasion, however, what NASA had to say was amazing:

A primitive ocean on Mars held more water than Earth's Arctic Ocean, according to NASA scientists who, using ground-based observatories, measured water signatures in the Red Planet's atmosphere. "Our study provides a solid estimate of how much water Mars once had, by determining how much water was lost to space," said Geronimo Villanueva, a scientist at NASA's Goddard Space Flight Center in Greenbelt, Maryland, and lead author of the new paper. "With this work, we can better understand the history of water on Mars." Perhaps about 4.3 billion years ago, Mars would have had enough water to cover its entire surface in a liquid layer about 450 feet (137 meters) deep. More likely, the water would have formed an ocean occupying almost half of Mars' northern hemisphere, in some regions reaching depths greater than a mile (1.6 kilometers).[7]

NASA explained:

From the measurements of atmospheric water in the near-polar region, the researchers determined the enrichment, or relative amounts of the two types of water, in the planet's permanent ice caps. The enrichment of the ice caps told them how much water Mars must have lost—a volume 6.5 times larger than the volume in the polar caps now. That means the volume of Mars' early ocean must have been at least 20 million cubic kilometers (5 million cubic miles).[8]

In conclusion, what we are seeing here—and what NASA has been able to confirm across the course of no less than several decades—is the existence of a huge body of data that offers us an amazing scenario: Mars was once a diverse world, one that perhaps was not at all too different to ours—a world that was filled with oceans (or, at the very least, with one), lakes, and rivers. When we add all of that to the pictures of those huge, enigmatic structures that pepper the landscape, to the photographs of highly unusual artifacts on the Martian floor, and maybe even to images of Martian animals, what we have—or what we *should* have—is a new appreciation for the Red Planet, a place that is seemingly forever dishing up new wonders and new mysteries to be solved. Or, in some cases, to be deliberately hidden. Collectively, these combined wonders create an amazing image of what life just might have been like millions of years ago—perhaps even longer ago than that. On the other hand, all that we have learned in this chapter provokes a distinctly uneasy question: What could have led such a beautiful world to have suffered from almost total destruction?

We'll soon get to that.

CHAPTER 18

"The Legend of Levitation"

|||||||||||||||||||||||||||||||||||||||

If there is one thing that we can say for sure about the mysterious formations and constructions on Mars, it's that they are huge. For example, the Face on Mars is approximately two kilometers in length, and the D&M Pyramid is more than one kilometer (approximately .6 mile) in height. Consider the fact that the Great Pyramid of Giza is "only" 481 feet in height, and you will have some appreciation of the sheer, massive nature of just some of the Martian anomalies. If the D&M Pyramid, in particular, was the work of ancient Martians, then it surely begs a most important question: How was such an immense structure created? How was the Face on Mars altered from a natural mesa to something much more? The answer may well be this: by the very same technique that was possibly used to erect some of the more famous creations on our world in the distant past, such as Stonehenge and the pyramids of Egypt. It's time to become acquainted with what is known as *acoustic levitation*. Throughout history, folklore, and mystery, tales have circulated of massive stones being moved through the air—effortlessly through the air, I should add—via sound. Bizarre? Well, yes, it is. That doesn't take away the fact that it's likely that sound was, and still remains, the key to the construction of the pyramids of Egypt, Stonehenge, and the massive stones at Baalbek, Lebanon.

Two of the leading figures in the research of acoustic levitation are Marie Jones and Larry Flaxman. They explain acoustic levitation in easy to understand, concise fashion: "Quite simply, an acoustic levitator is nothing more than a resonance machine of sorts—a way of introducing two opposing sound frequencies with interfering sound waves, thus creating a resonant zone that allows the levitation to occur. Theoretically, to move a levitating object, simply change or alter the two sound waves and tweak accordingly."[1]

Was this the technology that the Martians employed in such an undeniably spectacular fashion across Cydonia and elsewhere? The likelihood, as I see it, is very high. Let's see what we know of the old-but-intriguing accounts of how sound was utilized to make massive stones move near-effortlessly through the air.

Morris K. Jessup was someone who spent much of the 1950s investigating the legends of ancient mysteries and levitation. His books included *The Case for the UFO, UFOs and the Bible,* and *The Expanding Case for the UFO.* Jessup noted that: "The Black Pagoda of India, believed by some to have been seven centuries ago, is 228 feet high. Its roof, twenty-five feet thick, is a single stone slab weighing 2,000 tons" and added: "Levitating it into place seems as credible to me as any known mechanism including a sand ramp. Even then I do not see how they could possibly get sufficient purchase on this thing to move it at all, or attach sufficient harness to apply the necessary force."[2]

Jessup was most certainly no armchair researcher: He spent lengthy periods of time traveling around Guatemala, Belize, Mexico, and South America. He was particularly fascinated by the incredible stonework of the Incas and the Mayas. It was this fascination that led Jessup to take a trip to Uxmal. UNESCO [the United Nations Educational, Scientific and Cultural Organization] says of Uxmal:

The Mayan town of Uxmal, in Yucatán, was founded c. A.D. 700 and had some 25,000 inhabitants. The layout of the buildings, which date from between 700 and 1000, reveals a knowledge of astronomy. The Pyramid of the Soothsayer, as the Spaniards called it, dominates the ceremonial center, which has well-designed build-ings decorated with a profusion of symbolic motifs and sculptures depicting Chaac, the god of rain. The ceremonial sites of Uxmal, Kabah, Labna and Sayil are considered the high points of Mayan art and architecture.[3]

Uxmal has a fascinating legend attached to it that fits in neatly with the material contained in this particular chapter. The history of the Mayans states that its origins date back to around 500 AD, with significant building taking place in the ninth and tenth cen-turies. By the sixteenth century, however, Uxmal was a dead place, its people having abandoned it in widespread fashion. There is, however, an eye-opening story that revolves around what is called the Pyramid of the Magician—also the Pyramid of the Soothsayer and the Pyramid of the Dwarf. It's a Mesoamerican step pyramid that sits in the heart of the old Mayan ruins, and is surrounded by lush, green trees.

There is a notable reason why one of the several names of the pyramid is the Pyramid of the Dwarf. Legend has it that a mysteri-ous, small humanoid (or, perhaps, humanoids) oversaw the con-struction of the 131-foot-high pyramid. In addition, the creature was said to have been able to raise and direct the stones into place via a very novel and alternative fashion—namely, by whistling in the direction of the stones, something that caused them to rise up, and allowing them to be put into place with an absolute minimum of effort. We should not assume that one can move vast stones through the air just by whistling at them. Without a doubt that's absurd. What is not so absurd, however, is how acoustic levitation can lift objects—and how it involves sound. It's entirely plausible that the Mayans knew that sound played a role in the stone-raising.

And, as a result, and over a fairly significant period of time, they turned the science of acoustic levitation into simplistic whistling in their legends and folklore.

On top of all that, Morris Jessup had what can only be described as an obsession with the mysteries of Baalbek: It is situated near the Litani River in Lebanon's Beqaa Valley and is home to what is known as the Stone of the Pregnant Woman. This gigantic slab of stone weighs in at an incredible 1,000.12 tons. A second huge block in the area is 240 tons heavier than the more famous Stone of the Pregnant Woman. Matters don't end there, though: In 2014, yet another huge stone was unearthed. This one was partially buried beneath the Stone of the Pregnant Woman. It was the absolute granddaddy of all the stones, weighing *1,650 tons*. Moving such massive, heavy stones across desert landscapes, using rollers, rope, and manpower practically begs belief.

And there is more, too, on the matter of mystifying levitation in the past. In 1974, researcher-author Richard Mooney, the author of *Colony Earth*, said: "There is a tradition that appears in the mythology of the Americas that the priests 'made the stones light,' so that they were moved easily. This connects with the legend of levitation, which may have referred originally to an actual technique or device, long since forgotten."[4]

Now, let us address the claims of one Abu al-Hasan Ali al-Masudi and his connection to the pyramids of Egypt and matters relative to levitation. Also referred to as the "Herodotus of the Arabs," al-Masudi was born in Baghdad in the ninth century and was the author of *The Meadows of Gold and Mines of Gems*, a book that ran into a number of volumes and that chronicled his many expeditions and what, today, we would call road trips. Al-Masudi's travels were impressive, in terms of their scope: Egypt, India, Armenia, and Syria were just some of the lands that he visited. It was while on his travels that Baghdad-born al-Masudi came across an incredible story concerning the construction of the Giza Necropolis.

Forget those aforementioned rollers and ropes. So the story told to al-Masudi goes, the creators of the pyramids had in their possession what was described as a mysterious papyrus that could work wonders when it came to the matter of shifting stones—and that's an understatement of epic proportions. The first part of the operation was to place the papyrus below the corners of the massive blocks. Then, an equally mysterious rod of metal would be used by the builders to tap the stones one-by-one—something that apparently caused the stones to slowly and seemingly effortlessly rise up from the ground. One-by-one, and in single-file fashion, the stones were raised to approximately the height of a man, thus allowing the builders to push the stones along, as easily as one might push a child's toy boat on a body of water. After a distance of somewhere in the region of about 150 feet, the stones would slowly float back to the ground and the process would have to be repeated again—and again and again—until the stones were firmly in place, as a result of carefully directing them through the sky and to their final resting place. To be sure, it would have been an incredible sight.

It's important to note that, regardless of whether you are of the opinion that acoustic levitation is the answer to this riddle, it is a fact that the Egyptians had a deep knowledge of sound and its importance. Andrew Collins, the author of many books, including *Beneath the Pyramids* and *Gods of Eden*, offers these words concerning the Great Pyramid: "Precision geometry incorporating harmonics, proportions, and sound acoustics was incorporated into its exterior and interior design." He adds: "It has long been known that many of the temples and monuments of Pharaonic Egypt incorporate an intimate knowledge of sound acoustics." And most notable of all, Collins said that he had uncovered data showing that, as late as the 1900s, certain monasteries in Tibet harbored "a sonic technology that included the creation of weightlessness in stone blocks."[5]

And let us not forget that sound is alleged to have played a role in the raising of the blocks at Stonehenge: Legendary giants employed music to levitate the stones. This brings us to a March 2014 BBC article titled "Stonehenge Bluestones Had Acoustic Properties, Study Shows." A portion of the article makes for particularly intriguing reading:

> The giant bluestones of Stonehenge may have been chosen because of their acoustic properties, claim researchers. A study shows rocks in the Preseli Hills, the Pembrokeshire source of part of the monument, have a sonic property. Researcher Paul Devereux said: "It hasn't been considered until now that sound might have been a factor." With this study, thousands of stones along the Carn Menyn ridge were tested and a high proportion of them were found to "ring" when they were struck. "The percentage of the rocks on the Carn Menyn ridge are ringing rocks, they ring just like a bell," said Mr. Devereux, the principal investigator on the Landscape and Perception Project. "And there's lots of different tones, you could play a tune. In fact, we have had percussionists who have played proper percussion pieces off the rocks."[6]

As I said earlier, we should not take it literally that music and whistling somehow raised stones the weights of dozens of modern-day cars into the skies above. Such a thing is not possible. But, music and whistling have one thing in common: sound. *Acoustics.* Almost certainly, acoustic levitation was at the heart of these incredible feats. No doubt, and as the centuries passed, the truth of the science behind acoustic levitation was lost and forgotten—and distorted, too—with little more left than fanciful tales of music, whistles, curious papyrus, and strange metal rods that could achieve incredible feats in the air. Today, we are finally starting to get a grasp on this incredible technology, albeit still on a very small scale, as Marie D. Jones and Larry Flaxman reveal: "Scientists have successfully used the properties of sound to cause heavy gases, liquids, and even solid objects (such as spiders, goldfish, and mice) to float in the air."[7]

Moving on to the next chapter, if Ingo Swann was correct regarding his remote viewing–based conclusions—that a significant number of Martians successfully fled their doomed world and made it to the Moon and the Earth—then this requires an important and mystery-filled question to be answered: What caused the near-destruction of Mars and its lifeforms? It's now time to address matters relative to nuclear warfare, disaster in the Asteroid Belt, and a near-destroyed atmosphere on the Red Planet.

CHAPTER 19

"Ancient Martian Civilization Was Wiped Out"

||||||||||||||||||||||||||||||||||||||

On the matter of theoretical ancient nuclear war on Mars in an era long ago, one of the most important statements of all comes from Vince DiPietro, who, with Gregory Molenaar, pretty much began the Face on Mars–based research all on their own, in the 1970s. His words were driven by Dr. John Brandenburg, a man whom we will come to imminently. DiPietro said, in a 1998 paper titled "Mars: The Planet of Mysteries," that:

> The element Xenon 129, the second generation of a radioactive component produced when nuclear fission occurs, is found in abundance in the Martian atmosphere. Nuclear fission, such as that from a reactor or bomb, produces Iodine 129 with a half life of 17 million years, releasing beta particles and Xenon 129. The latter element is stable and lasts forever. Has nuclear fission taken place on Mars?[1]

It's an incredibly important question that needs to be answered, as it gets to the heart of the possibility that the Martians may have all but destroyed themselves while we were still living in caves and racing around and looking for the occasional wooly mammoth to dine upon.

As Dr. Brandenburg's bio reveals:

John E. Brandenburg is a theoretical plasma physicist who was born in Rochester, Minnesota, and grew up in Medford Oregon. He obtained his BA in Physics, with a Mathematics minor, from Southern Oregon University in 1975 and obtained his MS in 1977 and PhD in Plasma Physics both from University of California at Davis in 1981. He presently is working as a consultant at Morningstar Applied Physics LLC and a part-time instructor of Astronomy, Physics and Mathematics at Madison College and other learning institutions in Madison Wisconsin. Before this, he worked at Orbital Technologies in Madison Wisconsin, as Senior Propulsion Scientist, working on space plasma technologies, nuclear fusion, and advanced space propulsion.[2]

Impressive, indeed.

And, Brandenburg is at the forefront of the research to determine if a Martian race—or races, even—exterminated itself in a nuclear holocaust. In 2014, the world's media provided significant coverage to Brandenburg's important work. On November 21st of that year the UK's popular *Daily Mail* ran an article that was all but certain to turn heads. The title was: "Ancient Martian Civilization Was Wiped out by Nuclear Bomb-Wielding Aliens—and They Could Attack Earth Next, Claims Physicist."

Jonathan O'Callaghan, the journalist who wrote the feature for the *Daily Mail,* said:

If you're planning to go to the 2014 Annual Fall Meeting of the American Physical Society in Illinois this Saturday, you might be in for a bit of a surprise with the final talk of the day. Because that's when plasma physicist Dr. John Brandenburg will present his theory that an ancient civilization on Mars was wiped out by a nuclear attack from another alien race. In his bizarre theory, Dr. Brandenburg says ancient Martians known as Cydonians and Utopians were massacred in the attack—and evidence of the genocide can still be seen today.[3]

Brandenburg himself said:

The Martian surface is covered with a thin layer of radioactive substances including uranium, thorium and radioactive potassium—and this pattern radiates from a hot spot on Mars. A nuclear explosion could have sent debris all around the planet. . . . Taken together, the data requires that the hypothesis of Mars as the site of an ancient planetary nuclear massacre, must now be considered.[4]

Mac Tonnies had his suspicions that planet-wide altercations of the nuclear kind may have been at the heart of the puzzle of what happened to Mars. He said: "The D&M Pyramid is swollen and cracked, as if once molten. Despite this, no signs of volcanism are apparent. An unknown dark, sooty material has settled into fine-scale fractures, with a thick concentration near what researchers have referred to as a 'domed uplift,' thought by some to represent *an ancient internal explosion* [author's emphasis]."[5]

He also felt that it was important enough to highlight "early speculation by Richard Hoagland and John Brandenburg [who] raised the possibility that the domed uplift . . . is the result of 'explosive penetration.'"

The story, for Tonnies, was far from over. He posed a thought-provoking question that really caused controversy: "Could the D&M Pyramid have been deliberately destroyed from within by some ancient act of sabotage or act of war?"

Tonnies provided a potential answer to his question: "Branden-burg, a plasma physicist and coauthor, with Monica Rix Paxson, of *Dead Mars, Dying Earth,* estimates that *a one-ton kiloton nuclear explosion* [author's emphasis] could account for the damage visibly seen . . . assuming the deforming event was of an artificial nature."

"Unlike other formations in Cydonia," Tonnies carried on to me, "the D&M shows evidence of *having been melted* [author's emphasis]. The terrain around the apparent fifth buttress is cha-otic and spotted with enigmatic dark areas. . . . Additionally, Mark Carlotto has noted that there appears to be a dark material oozing out of the area associated with the domed uplift. This is consistent with an internal explosion."[6]

Bruce Rux, author of *Architects of the Underworld,* a book that is filled with Mars-based mysteries, says: ". . . the substantial apparent impact damage on many of Mars' surface structures, would lend credence to the biblical 'war in the heavens.' These include an extremely dramatic hole, one thousand feet in diameter, in Cydo-nia's D&M Pyramid, with structural and surface damage and debris, looking very much like it was caused by explosive penetration."[7] In addition, Rux suggests that the huge pyramid displays evidence of "clear impacts on the pyramid, with ejecta, high toward the top and lower down on the flank of its eastern face; the violence of the higher impact had thrown debris clearly over onto the western face."[8]

Certainly, there were equally apocalyptical—but very different—theories for what may have caused the wipeout of the Martians, as Mac Tonnies acknowledged to me:

Astronomer Tom Van Flandern has proposed that Mars was once the moon of a tenth planet that literally exploded in the distant past. If so, then the explosion would have had severe effects on Mars, probably rendering it uninhabitable. That's one rather apoc-alyptic scenario. Another is that Mars' atmosphere was destroyed by the impact that produced the immense Hellas Basin. Both ideas are fairly heretical by current standards; mainstream planetary science is much more comfortable with Mars dying a slow, prolonged death. Pyrotechnic collisions simply aren't intellectually fashionable—despite evidence that such things are much more commonplace than we'd prefer.[9]

Did the Martian civilization all but destroy itself in a terrible nuclear war in the distant past, as John Brandenburg has skillfully and logically suggested? If that was the case, it's not at all implausible that the Martians fled their world for ours, as we have seen—and maybe even created some form of home for themselves on the Moon, too. It may very well also be the case that the warlike Martians, after settling on Earth, scarred our landscape in just the way they did to their world—though not on a planet-wide scale, thankfully. We're talking about tactical nuclear weapons and the obliteration of entire cities and their people. We may never know for sure what or who it was that provoked nuclear exchanges on our world millennia ago; what we can say, however, is that evidence most assuredly does exist to show that apocalyptic-like events did occur. If the Martians created havoc and destruction on our planet, then the chances are that we will never be able to prove it. After all, we are talking about periods in the Earth's history when human civilization was still in a primitive state, and, as a result, we are left with little to nothing in the way of priceless and precious documentation chronicling such calamities. In terms of events that occurred relatively recently in our history, however, we are on far more solid ground. We'll begin with the Holy Bible's Old Testament saga of Sodom and Gomorrah.

Although we don't have a solid date for when the two cities were vaporized in the blink of an eye, the consensus is that the destruction occurred at some point around 2000 BC. It's a story that appears in the Qur'an, the Hadith, and the Old Testament. This is important, because it provides us with no less than three sources for the story (something that certainly helps us when it comes to the matter of corroboration). As to where Sodom and Gomorrah were located, it's believed that they were constructed on the edges of the River Jordan. While the tale of the obliteration of the two cities is very well known, far less well known is the fact that the overlords of Sodom and Gomorrah made an alliance with the rulers of other cities in the area—cities that were also wiped off the face of the map. They were Zeboim, Admah, and Zoar. According to the legends, it was only the people of Zoar who were not annihilated on the spot.

So the story goes, the massive destruction was provoked by what was termed the War of Nine Kings. The old texts reveal that for decades a powerful king in the area, Chedorlaomer, ruled ruthlessly throughout the land, something that led to a conflict between what were known as the Four Southern Kings and the Seven Northern Kings. The Old Testament, specifically Genesis 14, tells of the ominous countdown that led to destruction and death on a massive scale.

> At the time when Amraphel was king of Shinar, Arioch king of Ellasar, Kedorlaomer king of Elam and Tidal king of Goyim, these kings went to war against Bera king of Sodom, Birsha king of Gomorrah, Shinab king of Admah, Shemeber king of Zeboyim, and the king of Bela (that is, Zoar). All these latter kings joined forces in the Valley of Siddim (that is, the Dead Sea Valley). For twelve years they had been subject to Kedorlaomer, but in the thirteenth year they rebelled.
>
> In the fourteenth year, Kedorlaomer and the kings allied with him went out and defeated the Rephaites in Ashteroth Karnaim, the Zuzites in Ham, the Emites in Shaveh Kiriathaim and the Horites in the hill country of Seir, as far as El Paran near the desert. Then they turned back and went to En Mishpat (that is, Kadesh), and they conquered the whole territory of the Amalekites, as well as the Amorites who were living in Hazezon Tamar.[10]

Then the king of Sodom, the king of Gomorrah, the king of Admah, the king of Zeboyim and the king of Bela (that is, Zoar) marched out and drew up their battle lines in the Valley of Siddim against Kedorlaomer king of Elam, Tidal king of Goyim, Amraphel king of Shinar and Arioch king of Ellasar—four kings against five. Now the Valley of Siddim was full of tar pits, and when the kings of Sodom and Gomorrah fled, some of the men fell into them and the rest fled to the hills. The four kings seized all the goods of Sodom and Gomorrah and all their food; then they went away. They also carried off Abram's nephew Lot and his possessions, since he was living in Sodom.

A man who had escaped came and reported this to Abram the Hebrew. Now Abram was living near the great trees of Mamre the Amorite, a brother of Eshkol and Aner, all of whom were allied with Abram. When Abram heard that his relative had been taken captive, he called out the 318 trained men born in his household and went in pursuit as far as Dan. During the night Abram divided his men to attack them and he routed them, pursuing them as far as Hobah, north of Damascus. He recovered all the goods and brought back his relative Lot and his possessions, together with the women and the other people.

After Abram returned from defeating Kedorlaomer and the kings allied with him, the king of Sodom came out to meet him in the Valley of Shaveh (that is, the King's Valley).

Then Melchizedek king of Salem brought out bread and wine. He was priest of God Most High, and he blessed Abram, saying,

"Blessed be Abram by God Most High,
Creator of heaven and earth.
And praise be to God Most High,
who delivered your enemies into your hand."

Then Abram gave him a tenth of everything.

The king of Sodom said to Abram, "Give me the people and keep the goods for yourself."

But Abram said to the king of Sodom, "With raised hand I have sworn an oath to the Lord, God Most High, Creator of heaven and earth, that I will accept nothing belonging to you, not even a thread or the strap of a sandal, so that you will never be able to say, 'I made Abram rich.' I will accept nothing but what my men have eaten and the share that belongs to the men who went with me—to Aner, Eshkol and Mamre. Let them have their share."

Though the defeat of Chedorlaomer was seen as a good thing—which it was for those who lived in fear of him—perilous events were looming large on the horizon. The story tells of how God—or, maybe, a body of moralistic Martians—became more and more angered by what they saw as the lax morals of the people of Sodom and Gomorrah. And, remember, we are talking about the God of the Old Testament, who is presented in a very different way than God is described in the New Testament. The latter was a kindly figure, whereas the deity of the Old Testament was a dangerous and deadly figure who had no qualms about killing people on a large scale. And, we're talking about men, women, and children. Much of God's wrath was provoked by the sex lives of the people of Sodom and Gomorrah. Homosexuality was condemned by God, as was sodomy—which gets its terminology directly from the ancient city. Sex for money and what, today, we might call swinging lifestyles were also on God's "list." He had just about enough of what he saw as the abominable lives of the cities' people. Thankfully, today we live in far more enlightened times compared to those of thousands of years ago. Now, we get to the really intriguing part of the story.

According to the Old Testament, it wasn't long before Sodom and Gomorrah were destroyed that three strange characters stealthily approached Abraham, one of the formative figures in the Holy Bible. Angels taking on the appearance of people—that's the overriding theory. The meeting between humans and angels occurred on the Plains of Mamre. Abraham and Sarah (the latter being Abraham's wife) were given the bad news—the very bad news. In fact, it was just about the worst news possible: God was going to kill all of the people of both cities, as a result of their lives of wild debauchery. A panicked Abraham realized that he, his loved ones, and his family and friends would soon be no more. He quickly came up with a plan. Abraham put a question to the angelic trio: If he (Abraham) was able to find fifty men in the cities who had not succumbed to what God saw as wholly unacceptable behavior, could the cities and their people be spared? The answer, to Abraham's absolute relief, from God was *yes*.

Unfortunately, things didn't work out in the fashion that Abraham had hoped.

It turns out that Abraham was not the only one who was visited. Two angelic entities soon knocked on the door of the home of Lot, who was Abraham's nephew and one of the central players in the story of Sodom and Gomorrah and the downfall of the cities. Lot and his family were given the grim news. Lot took it as well as he could, and invited the mysterious pair to not just dine with him and his family, but to spend the night. That's where things went drastically wrong, as we learn in Genesis 19:4–5 (KJV): "But before they lay down, the men of the city, even the men of Sodom, compassed the house round, both old and young, all the people from every quarter: And they called unto Lot, and said unto him, Where are the men which came in to thee this night? Bring them out unto us, that we may know them."

It quickly became obvious that the men who soon surrounded Lot's house were determined to have sex with the two angels—and clearly by force. Lot, in a state of fear, however, begged them to change their minds regarding this "wicked thing." Lot was in such a state of panic that he even resorted to offering his two daughters, both virgins, to the men instead. They would not listen in the slightest, something that set the countdown to destruction in what turned out to be irreversible motion. When the men of the city tried to force their way into Lot's house, they were—rather intriguingly—blasted with a device that blinded them, something that brought the chaos to a sudden halt.

Despite the plan of Abraham's and Lot's bargaining of the very controversial kind, God was having no more. He was done with Sodom and Gomorrah. The cities and their people would soon be gone—completely eradicated from the face of the Earth. There was a small piece of mercy, however: The angels confided in Lot what was looming large on the horizon—namely, death on a massive scale, and Sodom and Gomorrah turned to dust. It was, however, just enough time for Lot, his wife, and their daughters to flee their doomed home. Of this development, Genesis (KJV) states:

> The two men said to Lot, "Do you have anyone else here—sons-in-law, sons or daughters, or anyone else in the city who belongs to you? Get them out of here, because we are going to destroy this place. The outcry to the Lord against its people is so great that he has sent us to destroy it."
>
> So Lot went out and spoke to his sons-in-law, who were pledged to marry his daughters. He said, "Hurry and get out of this place, because the Lord is about to destroy the city!" But his sons-in-law thought he was joking.
>
> With the coming of dawn, the angels urged Lot, saying, "Hurry! Take your wife and your two daughters who are here, or you will be swept away when the city is punished." When he hesitated, the men grasped his hand and the hands of his wife and of his two daughters and led them safely out of the city, for the Lord was merciful to them. As soon as they had brought them out, one of them said, "Flee for your lives! Don't look back, and don't stop anywhere in the plain! Flee to the mountains or you will be swept away!" But Lot said to them, "No, my lords, please! Your servant has found favor in your eyes, and you have shown great kindness to me in sparing my life. But I can't flee to the mountains; this disaster will overtake me, and I'll die. Look, here is a town near enough to run to, and it is small. Let me flee to it—it is very small, isn't it? Then my life will be spared."
>
> He said to him, "Very well, I will grant this request too; I will not

overthrow the town you speak of. But flee there quickly, because I cannot do anything until you reach it."

Genesis (KJV) tells of how the nightmare began and how it ended. For Lot's wife it ended badly, *very* badly:

> By the time Lot reached Zoar, the sun had risen over the land. Then the Lord rained down burning sulfur on Sodom and Gomorrah—from the Lord out of the heavens. Thus he overthrew those cities and the entire plain, destroying all those living in the cities—and also the vegetation in the land. But Lot's wife looked back, and she became a pillar of salt.
>
> Early the next morning Abraham got up and returned to the place where he had stood before the Lord. He looked down toward Sodom and Gomorrah, toward all the land of the plain, and he saw dense smoke rising from the land, like smoke from a furnace.
>
> So when God destroyed the cities of the plain, he remembered Abraham, and he brought Lot out of the catastrophe that overthrew the cities where Lot had lived.[11]

Lot was not happy with Zoar, and so he and his daughters did as they were told: They raced to the safety of the mountains, which were peppered with caves and provided much-needed cover. It was while the three were deep in the caves that something controversial happened—extremely controversial: The two girls got their father drunk and had sex with him. Supposedly, the girls' plan was to "preserve our family line through our father." The Old Testament states: "So both of Lot's daughters became pregnant by their father. The older daughter had a son, and she named him Moab; he is the father of the Moabites of today. The younger daughter also had a son, and she named him Ben-Ammi; he is the father of the Ammonites of today" (Genesis 19:36 NIV).

That, in essence, is the story of Sodom and Gomorrah—and of Lot and his family. The tale is not quite over, however. A careful look at the tale allows us to make a strong case that the cities were destroyed by tactical nuclear weapons.

There was a fascinating development in the story of Sodom and Gomorrah in 2008. It revolved around the work of one Sir Austen Henry Layard, an archaeologist, who died in 1894 at the age of 77. While digging in Nineveh, an ancient Mesopotamian city that stood in what is now Northern Iraq, Layard came across an incredibly old planisphere—a clay tablet, in simple terminology. Of the developments in 2008, the *Register* stated the following:

Alan Bond, Managing Director of Reaction Engines Ltd and Mark Hempsell, Senior Lecturer in Astronautics at Bristol University, subjected the Planisphere to a program which "can simulate trajectories and reconstruct the night sky thousands of years ago." They discovered that it described "events in the sky before dawn on the 29 June 3123 BC," with half of it noting "planet positions and cloud cover, the same as any other night." The other half, however, records an object "large enough for its shape to be noted even though it is still in space" and tracks its trajectory relative to the stars, which "to an error better than one degree is consistent with an impact at Köfels."[12]

As for what might have come down, the team suggested it could have been a gigantic asteroid. Bristol University (UK) got involved in the revelations:

The observation suggests the asteroid is over a kilometer in diameter and the original orbit about the Sun was an Aten type, a class of asteroid that orbit close to the earth, that is resonant with the Earth's orbit. This trajectory explains why there is no crater at Köfels. The incoming angle was very low (six degrees) and means the asteroid clipped a mountain called Gamskogel above the town of Längenfeld, 11 kilometers from Köfels, and this caused the asteroid to explode before it reached its final impact point. As it travelled down the valley it became a fireball, around five kilometers in diameter (the size of the landslide). When it hit Köfels it created enormous pressures that pulverized the rock and caused the landslide but because it was no longer a solid object it did not create a classic impact crater.[13]

The following was added: "Another conclusion can be made from the trajectory. The back plume from the explosion (the mushroom cloud) would be bent over the Mediterranean Sea re-entering the atmosphere over the Levant, Sinai, and Northern Egypt. The ground heating though very short would be enough to ignite any flammable material—including human hair and clothes. It is probable more people died under the plume than in the Alps due to the impact blast."[14]

The prime problem in this admittedly intriguing development is that the Köfels incident happened about 10,000 years ago. This is long before the destruction of Sodom and Gomorrah were destroyed—unless, that is, we need to radically alter the timeframe when the two cities were eradicated. It should be noted that the news of 2008 had a precursor to it.

Back in 1997, staff at the UK's Cambridge University put together a seminar to discuss data that "provides dramatic evidence for an extraterrestrial cause for the wholesale collapse of several civilizations around 2200 BC."[15] No prizes for guessing that this "wholescale collapse" included Sodom and Gomorrah. For the staff at Cambridge, the likely candidate was not an asteroid, but—probably—a comet.

The *Independent*'s environment correspondent, Geoffrey Lean, wrote:

The conference, on natural catastrophes during Bronze Age civilizations, will bring together astronomers, archaeologists, geologists and other scientists to try to find an explanation for the near-simultaneous fall of the Old Kingdom of ancient Egypt, the Sumerian civilization in Mesopotamia and the Harrapin Civilization of the Indus Valley. In all, some 40 cities are thought to have disappeared, in a series of catastrophes.[16]

This is all perfectly logical and plausible. After all, for most people a comet or an asteroid is probably a far more likely candidate than ancient Martians, with strict morals, deciding to annihilate who knows how many people with nuclear weapons. There is, however, an important issue to remember and keep firmly in mind: Lot and his family were not just warned of when the overwhelming destruction would occur. As the old story goes, they were warned *ahead* of the flattening of Sodom and Gomorrah.

If the words of the Old Testament can be viewed as being accurate, then the comet and asteroid theories fall down flat. After all, predicting the *exact* impact sites of comets, meteorites, and asteroids, today even, is not an easy task; never mind way back then. Yet, we're told that thousands of years ago, Lot knew exactly when and where the holocaust would hit. The only feasible answer to this particular conundrum is that, whatever it was that hit Sodom and Gomorrah, it was specifically directed to the sites on a date and time that was shared with Lot, in advance of the chaos.

There is also the matter of those "angels" who warned Lot and his family to never look back, after heading out of the city that would soon be no more. There are a number of hazards from being in close proximity to a nuclear explosion that one needs to be aware of: burns, radiation exposure, and ruptured eardrums. And, there's the fact that—as the events that occurred in 1945 in the cities of Hiroshima and Nagasaki showed to a graphic degree—looking in the direction of a nuclear explosion can melt one's eyeballs. Literally. At the very least, temporary blindness can occur. At worst, permanent blindness can be the outcome. Those mysterious angels gave Lot vitally important information that helped him to survive intact when the terrifying, miles-high mushroom cloud suddenly dominated what was almost immediately a burning, melting, rumbling landscape.

As for the matter of those same angels blinding the crazed people who tried to invade Lot's house, well, such technology is in our hands today. So, why shouldn't ancient Martians have had similar weaponry way back then? On this matter, the Human Rights Watch Arms Project (HRWAP) says:

The United States has pursued the development of at least ten different tactical laser weapons that have the potential of blinding individuals. The existence of most of these programs is not known to the American public or to most of the U.S. Congress. In fact, the programs are little known even within the U.S. military, and services responsible for laser weapons seem largely unaware of the programs in research and development in other services. Further, the Office of the Secretary of Defense does not appear to have an overview of the program. Secrecy and lack of oversight and coordination are thus the hallmarks of the "family" of U.S. tactical laser weapons.[17]

The HRWAP also note:

The Air Force's Phillips Laboratory in Albuquerque, New Mexico developed Saber 203, "a laser system that can temporarily blind or impair the vision of enemy soldiers, reducing their ability to fight." It is suspected as having been developed for use by special operations forces as well as by Air Force security police. The Human Rights Watch Arms Project has identified the Saber 203 as the system deployed to Somalia with the Marine Corps in early 1995, although it appears that the Saber 203 is now controlled by the USSOCOM.[18]

This sounds astonishingly like a modern-day equivalent of just the kind of weaponry that may have been employed to protect Lot's "angels" from attack, when they warned him of the disaster that was destined to soon hit one and all.

Finally, let's focus our attentions on the most memorable part of the story: that of Lot's wife being transformed into a pillar of salt when the apocalypse hit. The late Zecharia Sitchin had an answer for this:

> [T]he notion that Lot's wife was turned into a "pillar of salt" when she stayed back to watch what was happening, in fact meant "pillar of vapor" in the original Sumerian terminology. Since salt was obtained in Sumer from vapor-filled swamps, the original Sumerian term NI.MUR came to mean both "salt" and "vapor." Poor Lot's wife was vaporized, not turned into salt, by the nuclear blasts that caused the upheaval of the cities of the plain.[19]

In "The Effects of Nuclear Weapons," Russell Hoffman wrote:

> Those within approximately a six square mile area (for a 1 megaton blast) will indeed be close enough to "ground zero" to be killed by the gamma rays emitting from the blast itself. Ghostly shadows of these people will be formed on any concrete or stone that lies behind them, and they will be no more. They literally won't know what hit them, since they will be vaporized before the electrical signals from their sense organs can reach their brains.[20]

Now, it's time to take a look at a number of eerily similar events that also occurred thousands of years ago. They, too, may have been the ruthless and reckless work of warmongering Martians who were forced to live on our world, as a result of the irreversible destruction they wrought on their own planet.

CHAPTER 20

"Massive Species Extinctions"

||||||||||||||||||||||||||||||||||||||||

Did a race of Martians destroy their world in times long, long gone? If so, then how was Mars rendered just about uninhabitable? To answer that question, we need to take a close look at the potential effects that a planet-wide nuclear war would have done to Mars, its people, its culture, and its landscape. The best way to do that is to address the matter of what a nuclear war would do to our very own world. In the 1950s, US government propaganda experts created ridiculous films showing that just about all one had to do in the event of World War Three was to hide under the bed, or under one's school desk, stay there for a few hours, and then come out, and hope for the best. And, in the aftermath, start over again. It almost sounds cozy and comforting, right? Well, yes, that was the whole point and the goal of it all: to keep people from worrying about what the future might bring and, instead, have them continue with their lives. The fact is, though, that a nuclear war between the superpowers would have decimated most of the population in the northern hemisphere and left those in the southern hemisphere to deal with deadly radiation. Billions, rather than millions, would have died, and civilization would have been in absolute ruins. Civilization would have quickly spiraled into savagery, as the irradiated, starving, burned, and psychologically destroyed survivors sought and fought to stay alive.

The Martians go to war. (US Department of Energy)

In just a few years—certainly no more than a few decades—our civilization would be forgotten. Those born into the world of destruction would have no concept of technology, science, the Internet, television, and just about anything and everything we rely on today. They would almost certainly be unable to read or write. After all, they would hardly be priorities in an irradiated world in which survival was the only goal. Those who were born, let us say, twenty or thirty years after the war, would see the destruction all around them—destroyed cities, rusted vehicles, and mountains of skeletons—but they would have little understanding of what it was they were seeing. Also, language would almost certainly dwindle, as more and more of the original people who survived the war died, and there were fewer and fewer people around to continue the process of teaching the young to speak. It's not an exaggeration to say that within a century, our civilization would be over, and the Earth would still be in tatters. We're not in a position to see how Mars's nuclear war played out, but we can make a good assessment of the effects nuclear war would have on our planet today.

It was in August 1945 that the Japanese cities of Nagasaki and Hiroshima were destroyed by atomic bombs delivered from the skies above by the US military. No one—including the scientists who worked at Los Alamos on the top-secret Manhattan Project, which was the operation to develop and deploy atomic weapons—was entirely sure what would happen on the fateful day of August 6, 1945, when Hiroshima was targeted for destruction. They soon found out, however. Everyone knew that the explosion was going to cause massive destruction, but to what extent? Well, the answer would only come in the seconds and minutes after the bomb was dropped. The answer came.

Approximately one hundred and twenty thousand people in Hiroshima died when the bomb detonated.[1] Thousands more were critically injured, burned, and irradiated. As for the city, it was in radioactive ruins. When, three days later, Nagasaki was all but bombed out of existence, about eighty thousand people were killed—many instantly and turned to ash (shades of the story and fate of Lot's wife as her family fled their doomed city thousands of years ago). Overall, the collective number of deaths was close to a quarter of a million. And it should be noted that this massive destruction and loss of life was caused by the earliest, most primitive atomic weapons of the era.

Jay Bennett, writing for *Popular Mechanics,* concisely makes it clear that in a worldwide nuclear war today, the destruction would be on such a level that it's almost impossible to comprehend it:

"These are the only two nuclear weapons ever used in warfare, and let's hope it stays that way, because some of the nuclear weapons today are *more than 3,000 times as powerful as the bomb dropped on Hiroshima* [author's emphasis]."[2] Today, the United States and Russia both have about six thousand nuclear weapons—all of them capable of mass destruction to near-unbelievable levels. Add to that the roughly two hundred nukes that the UK possesses and the approximately two hundred sixty that China has in its stockpiles, and one does not need to be a genius to realize how a nuclear confrontation involving the major nations of our world would effectively end civilization in just a few minutes—never mind hours, days, weeks, or years.[3]

Now, let's see what effects such a catastrophic war would have on the Earth itself. In doing so, we may well be able to gauge how and why Mars was so massively destroyed in its very own face-to-face hostilities of the alien kind.

In the event of a nuclear war, today, between Russia and the United States—and with the UK, Australia, Iran, Pakistan, and India almost certainly dragged in—the result would be terrible in the extreme: billions dead and the Earth in a state of chaos in mere minutes. There is, however, another, equally deadly phenomenon. It's what is known as a nuclear winter. Mikhail Gorbachev, the last leader of the now-defunct Soviet Union, stated:

In the 1980s, you warned about the unprecedented dangers of nuclear weapons and took very daring steps to reverse the arms race. Models made by Russian and American scientists showed that a nuclear war would result in a nuclear winter that would be extremely destructive to all life on Earth; the knowledge of that was a great stimulus to us, to people of honor and morality, to act in that situation.[4]

This brings up the obvious questions: What, precisely, is a nuclear winter? And, how would such a phenomenon affect the scattered survivors of the war? *Encyclopedia Britannica* gives us the answer. It notes that in the wake of a Third World War, millions of tons of

smoke and soot would be shepherded by strong west-to-east winds until they would form a uniform belt of particles encircling the Northern Hemisphere from 30° to 60° latitude. These thick black clouds could block out all but a fraction of the Sun's light for a period as long as several weeks. Surface temperatures would plunge for a few weeks as a consequence, perhaps by as much as 11° to 22° C (20° to 40° F). The conditions of semidarkness, killing frosts, and subfreezing temperatures, combined with high doses of radiation from nuclear fallout, would interrupt plant photosynthesis and could thus destroy much of the Earth's vegetation and animal life.[5]

How Stuff Works reveals a dark and disturbing scenario, too:

[N]uclear winter presents the notion that post–World War III humanity might very well die with a whimper. Since the early 1980s, this scenario has permeated our most dismal visions of the future: Suddenly, the sky blazes with the radiance of a thousand suns. Millions of lives burn to ash and shadow. Finally, as nuclear firestorms incinerate cities and forests, torrents of smoke ascend into the atmosphere to entomb the planet in billowing, black clouds of ash. The result is noontime darkness, plummeting temperatures and the eventual death of life on planet Earth.[6]

In 1985, the US National Research Council published a ground-breaking report titled *The Effects on the Atmosphere of a Major Nuclear Exchange*. Its conclusions, prepared by the Committee on the Atmospheric Effects of Nuclear Explosions, were terrifying:

The realization that a nuclear exchange would be accompanied by the deposition into the atmosphere of particulate matter is not new. However, the suggestion that the associated attenuation of sunlight might be so extensive as to cause severe drops in surface air temperatures and other major climatic effects in areas that are far removed from target zones is of rather recent origin.[7]

The group also revealed:

[T]he massive species extinctions of 65 million years ago were part of the aftermath of the lofting of massive quantities of particulates resulting from the collision of a large meteor with the earth. The consequences of any such changes in atmospheric state would have to be added to the already sobering list of relatively well-understood consequences of nuclear war. . . . Long-term atmospheric consequences imply additional problems that are not easily mitigated by prior preparedness and that are not in harmony with any notion of rapid postwar restoration of social structure. They also create an entirely new threat to populations far removed from target areas, and suggest the possibility of additional major risks for any nation that itself initiates use of nuclear weapons, even if nuclear retaliation should somehow be limited.[8]

Dr. Alan Robock, a professor of climatology in the department of environmental sciences at Rutgers University, has thought-provoking words to say on all of this too:

A minor nuclear war (such as between India and Pakistan or in the Middle East), with each country using 50 Hiroshima-sized atom bombs as airbursts on urban areas, could produce climate change unprecedented in recorded human history. This is only 0.03% of the explosive power of the current global arsenal.[9]

He also says:

65,000,000 years ago an asteroid or comet smashed into the Earth in southern Mexico. The resulting dust cloud, mixed with smoke from fires, blocked out the Sun killing the dinosaurs, and started the age of mammals. This Cretaceous-Tertiary (K-T) extinction may have been exacerbated by massive volcanism in India at the same time. This teaches us that large amounts of aerosols in Earth's atmosphere have caused massive climate change and extinction of species. The difference with nuclear winter is that the K-T extinction could not have been prevented.[10]

The Federation of American Scientists are acutely aware of the massive, planet-changing threat that nuclear winter promises to deliver:

Even if some humans survive, there could still be permanent harm to human civilization. Small patches of survivors would be extremely vulnerable to subsequent disasters. They also could not keep up the massively complex civilization we enjoy today. It would be a long and uncertain rebuilding process and survivors might never get civilization back to where it is now. More importantly, they might never get civilization to where we now stand poised to take it in the future. Our potentially bright future could be forever dimmed.[11]

The Atomic Archive has its views on all of this, too. On the aftermath of a nuclear war, they state that:

[T]hick black clouds could block out all but a fraction of the sun's light for a period as long as several weeks. The conditions of semi-darkness, killing frosts, and subfreezing temperatures, combined with high doses of radiation from nuclear fallout, would interrupt plant photosynthesis and could thus destroy much of the Earth's vegetation and animal life. The extreme cold, high radiation levels, and the widespread destruction of industrial, medical, and transportation infrastructures along with food supplies and crops would trigger a massive death toll from starvation, exposure, and disease.[12]

What all of this tells us is that a full-scale, nuclear war cannot be won. The only result of a confrontation between the most powerful nations on the planet—the United States and Russia—would be complete disaster for everyone and everything on Earth. Add to that the various other nations that have nukes (the UK, France, Israel, North Korea, India, Pakistan, and China) and imagine the outcome if all of those nations found themselves quickly pulled into the war and, as a result, *everyone* launches. Billions of people die. Animals are wiped out on massive scales. The world's most famous, bustling cities are gone—obliterated. And, on top of that there are the struggles of the survivors to try and combat a nuclear winter, which will continue not for months, but years or even centuries.

When we see what such a war would really do to us as a species, and when we also understand how even the Earth itself might be irreversibly pulverized and damaged, it's not at all difficult to see how, in an era that is now long gone, the Martians could have almost wiped out their race and—in the process—transformed their world into something barren, harsh, and bleak, and with its history, society, and culture almost completely wiped out.

CHAPTER 21

"High-Velocity Impacts"

||||||||||||||||||||||||||||||||||||||||

Still on the matter of nuclear war on Mars and on Earth, it's a certainty that many of the craters that can be found peppered throughout the Earth's surface were the result of meteorite and asteroid strikes—perhaps even comets, too. There are, however, some such craters that are very difficult to explain in down-to-earth fashion. That's right: We're talking about the possibility of yet further evidence of ancient nuclear war fought by Martians—but right here on Earth. We'll begin with the genuinely weird story of what is known as the Lonar Crater. It can be found in Maharashtra, India. The locals aside, no one had seen the crater until 1823. Interestingly, NASA has taken an interest in the ancient crater. NASA states of the now-water-filled crater:

> India's Lonar Crater began causing confusion soon after it was identified in 1823 by a British officer named C. J. E. Alexander. Lonar Crater sits inside the Deccan Plateau—a massive plain of volcanic basalt rock leftover from eruptions some 65 million years ago. Its location in this basalt field suggested to some geologists that it was

a volcanic crater. Today, however, Lonar Crater is understood to result from a meteorite impact that occurred between 35,000 and 50,000 years ago.[1]

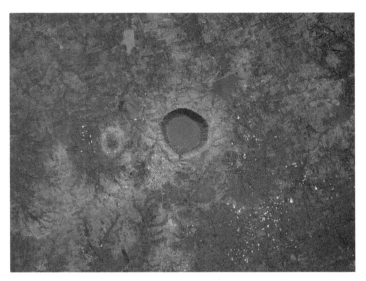

Destruction on Earth at the Lonar Crater. (NASA)

NASA continues:

Lonar Crater is approximately 150 meters (500 feet) deep, with an average diameter of almost 1,830 meters (6,000 feet). The crater rim rises roughly 20 meters (65 feet) above the surrounding land surface. Scientists established Lonar's status as an impact crater based on several lines of evidence, perhaps the most compelling being the presence of maskelynite. Maskelynite is a kind of naturally occurring glass that is only formed by extremely high-velocity impacts. A *Science* article published in 1973 pointed out this material's presence, and suggested that the crater's situation in volcanic basalt made it a good analogue for impact craters on the surface of the Moon.[2]

It should be noted that there is a very interesting, and intriguing, story attached to the history and lore of the Lonar Crater. For this part of the saga we have to turn our attentions to the Skanda-Purana, which is an approximately 1,100-year-old Hindu document, as Priya Ramachandran reveals:

> The name "Lonar" came from the demon, Lonasura, who lived in this subterranean abode and terrorized the people of the Earth. Now, heeding to his people's prayers Vishnu came to the rescue, by sending his avatar in the form of a man named Daityasudna, who exposed the demon's hideout, kicked away the rock that kept them hidden, thereby creating that crater. He slayed the mighty demon and when his blood was spilt, it turned into a lake.[3]

Nearby is the Daityasudan Temple. It was created as a place to pay homage to the defeat by Vishnu of the giant-sized Lonasura. Was Lonasura a surviving Martian? Was the "subterranean abode" in which he dwelled actually a highly sophisticated bunker, of the type that are in place today, to try and ensure at least some degree of survival in the event of a nuclear war? If Lonasura was indeed a Martian who made it to Earth—or, perhaps even, a Martian born on our world long after a Martian exodus from the Red Planet to our world occurred—one has to wonder if the massive crater was not created by the impact of a meteorite, but by a highly destructive device of the nuclear variety? On this matter, David Hatcher-Childress, who has extensively investigated the mystery makes a very good point:

> If such geologically recent craters as the Lonar [Crater] were of meteoric origin, why then do such tremendous meteorites not fall today? The Earth's atmosphere fifty thousand years ago probably was no different from today's, so a lighter atmosphere cannot be advanced as a hypothesis to explain an immense meteorite size, which of course would be considerably reduced by heat oxidization within a gaseously heavier atmosphere.[4]

Hatcher-Childress also states that "no trace of any meteoric, etc., material has been found at the site or in the vicinity, while it is the world's only known 'impact' crater in basalt. Indications of great shock (from a pressure exceeding six hundred atmospheres) and intense abrupt heat (indicated by basalt glass spherules) can be ascertained at the site."[5]

Now, let's address the matter of one of the world's most famous craters—or, arguably, *the* most famous one, period: Meteor Crater, located near Winslow, Arizona. This mammoth scarring of the landscape occurred around 50,000 years ago, and is more than two miles wide and more than five hundred feet deep. In an article titled "The First Global Nuclear War and a Coverup of HISTORICAL Proportions!" Tim Benedict made this thought-provoking statement:

> A large ground level, or even an underground "bunker busting," nuclear weapon would easily produce a crater consistent with the Arizona crater. If the site of the crater was originally a nuclear weapons production facility, it only makes sense that someone at war would want to bomb the site back to [the] stone-age, leaving behind a gaping crater that contains no actual meteorite fragments, but a surrounding area that is full of alloy, metallic, and other "meteor" fragments.[6]

With Benedict's words still in our minds, let us now address the deeper and stranger history of nuclear war in the United States thousands of years ago—wars fought by Martians who had, effectively, declared our world to be theirs.

Thousands of years ago, a massive, destructive event occurred in what is now the United States that wiped out entire swathes of the population. At the time, the dominant humans were the Clovis People, as they are known, and from whom today's Native Americans are descended. It's a baffling fact that when that aforementioned destructive event took place in North America, the number of people and animals that were obliterated was huge. Certainly, the Clovis People were most definitely not in a position to go to war with weapons of mass destruction; such a thing is utterly ridiculous. But, a case can be made that *someone*—possibly the survivors of a doomed Mars—unleashed incredibly powerful weaponry that flattened the landscape and ensured that life would never be quite the same again. Did various factions of the surviving Martians—and their descendants, even—go to war with each other on our planet, in much the same way they did on their own, original world? It's a controversial question that requires an answer.

In 2001, *Mammoth Trumpet Magazine* ran a feature written by Richard B. Firestone and William Topping that had an extremely eye-catching title: "Terrestrial Evidence of a Nuclear Catastrophe in Paleoindian Times." The pair made a careful and detailed study of what is known as the Great Lake Basin. It covers Michigan, Pennsylvania, Indiana, Ohio, Illinois, Wisconsin, New York, and Minnesota. As for the huge, famous bodies of water, they are Lake Ontario, Lake Huron, Lake Erie, Lake Superior, and Lake Michigan.

The pair wrote:

Our research indicates that the entire Great Lakes region (and beyond) was subjected to particle bombardment and a catastrophic nuclear irradiation that produced secondary thermal neutrons from cosmic ray interactions. The Paleoindian catastrophe was large by standards of all suspected cosmic occurrences. . . . The enormous energy released by the catastrophe at 12,500 yr B.P. could have heated the atmosphere to over 1000°C over Michigan, and the neutron flux at more northern locations would have melted considerable glacial ice. Radiation effects on plants and animals exposed to the cosmic rays would have been lethal, comparable to being irradiated in a 5-megawatt reactor more than 100 seconds.[7]

I should stress that Topping and Firestone were not suggesting that ancient aliens fought a war in the skies of what is now the

United States. Rather, their theory was that a supernova was the likely candidate.

Also on the trail for the truth of the mass extinction thousands of years ago were geophysicist Allen West and geologist James Kennett. While they, too, were sure that something from the skies had caused such widespread destruction, they suggested a comet, rather than a supernova, caused so much widespread apocalyptic activity. Their work continued for years. As an example of this, personnel at UC Santa Barbara in 2012 announced the following:

> An 18-member international team of researchers that includes James Kennett, professor of earth science at UC Santa Barbara, has discovered melt-glass material in a thin layer of sedimentary rock in Pennsylvania, South Carolina, and Syria. According to the researchers, the material—which dates back nearly 13,000 years— was formed at temperatures of 1,700 to 2,200 degrees Celsius . . . and is the result of a cosmic body impacting Earth.[8]

H. Richard Lane, of the National Science Foundation's Division of Earth Sciences, which came up with the much-needed dollars that allowed for the research to go ahead, stated:

> These scientists have identified three contemporaneous levels more than 12,000 years ago, on two continents yielding siliceous scoria-like objects. . . . SLO's are indicative of high-energy cosmic airbursts/impacts, bolstering the contention that these events induced the beginning of the Younger Dryas. That time was a major departure in biotic, human and climate history.[9]

Kennett and Lane had far more to say:

> If you imagine multiple nuclear explosions occurring over wide areas, generating major pressure waves, flash heat waves, knocking down forests. And this led to wildfires over wide areas, with major

destruction of the vegetation. The burning over broad areas of the continent would have destroyed the food resources for many of these animals. And, we suggest, that is why the larger animals, preferentially became extinct.[10]

Maybe the most intriguing words of Kennett were these:

The very high temperature melt-glass appears identical to that produced in known cosmic impact events such as Meteor Crater in Arizona, and the Australasian tektite field. *The melt material also matches melt-glass produced by the Trinity nuclear airburst of 1945 in Socorro, New Mexico* [author's emphasis]. The extreme temperatures required are equal to those of an atomic blast, high enough to make sand melt and boil.[11]

If you're wondering what the "Australasian tektite field" is, its formal title is the "Australasian strewn field." What it demonstrates is that close to 800,000 years ago much of our world was effectively torched. The editor of *Meteoritics and Planetary Science,* Stuart Ross Taylor, said:

Early workers, including such astute observers as Charles Fenner, George Baker, and Edward Gill, who picked up pristine tektites on the arid and ancient surface of Australia, became convinced that this shower of glass had arrived very recently. Ages around 10,000-20,000 years were usually quoted. As some early radiometric age determinations of tektites gave spuriously old ages, these data fueled the notion that the age of the fall was disconnected to the time of formation of tektites.[12]

And what are tektites? This is where things get *really* interesting. Tektites are small, glassy bodies that are near-identical to the equally glassy trinitite that was found at the detonation site of the first atomic bomb in New Mexico in July 1945, just a few weeks before the destruction of the Japanese cities of Nagasaki and Hiroshima.

The Los Alamos National Laboratory states:

Trinitite has also been referred to as atomsite or Alamogordo glass (after a nearby city). Ultimately, it is a glasslike substance that was created from the sand and other materials at the Site during the intense heat of that first atomic test. Relatively recent research (2005) indicates that, upon explosion, the ground was likely pushed down initially, then rebounded, forcing material into the fireball. As indicated in the display, the ground was vaporized before eventually raining down in the form of trinitite droplets. While most trinitite is light green (due to the iron that was present in the sand), other samples contain some of the iron from the tower on which the "Gadget" was detonated, and those look black. Yet other slightly red samples contain copper from the electrical wire used in the experiment.[13]

In 2001, *Science Frontiers* made that abundantly clear:

The extent of the immense Australasian-tektite strewn field implies a hard-to-miss crater about 100 kilometers in diameter. Yet, despite the geological recency of the event and despite much geological surveying, no convincing crater has been discovered. . . . The mystery deepens when one realizes that whatever cataclysm sent the Australian tektites aloft may have been comparable in magnitude to the impact that extinguished the dinosaurs (and other fauna) some 65 million years ago. This much older event has its crater buried below the Yucatan and is further marked by widespread biological extinctions. In contrast, the Australasian-tektite event is not only minus an obvious crater but seems to have had scant effect on the earth's cargo of sensitive life forms. It was a strangely "gentle" event despite the rocky deluge of tektites. What really happened?[14]

That's a very good question!

And, there's another relevant question that requires resolving: What about the stories of mysterious giants that surface now and again in the pages of this book? Maybe the Martians were not at all too dissimilar to us, except for one thing: they towered over us. As in literally.

CHAPTER 22

"The Reconstruction
of the
Devastated Earth"

IIIIIIIIIIIIIIIIIIIIIIIIIIIIIIIIIIIIIIIIII

The King James Version of the Bible tells us (Genesis 6:4): "There were giants in the earth in those days; and also after that, when the sons of God came in unto the daughters of men, and they bare children to them, the same became mighty men which were of old, men of renown."

Angela Sangster states of extraordinary tall people in ancient times:

> Were the giants spoken of in Genesis actually aliens from another planet? The *Enuma Elish* (the creation story according to the ancient Babylonians) has many parallels with the Christian bible. The *Book of Genesis* speaks of these giants (also called Nefilim) who were taught by the Church as having been angels who came to Earth and cohabitated with the daughters of Man (*Genesis* chapter six). *The Book of Enoch*, which was removed from the Christian *Bible*, speaks of the resulting hybrids of humans and these giants, or "watchers."[1]

Zecharia Sitchin, he of Anunnaki fame, and who we'll hear more from later, said: "Who were the *Nefilim*, that are mentioned in Genesis, Chapter six, as the sons of the gods who married the daughters of Man in the days before the great flood, the Deluge? The word Nefilim is commonly, or used to be, translated 'giants.'"

Sitchin added:

I questioned this interpretation as a child at school, and I was reprimanded for it because the teacher said "you don't question the *Bible*." But I did not question the *Bible*. I questioned an interpretation that seemed inaccurate, because the word, Nefilim, the name by which those extraordinary beings, "the sons of the gods" were known, means literally, "Those who have come down to earth from the heavens."[2]

The staff at Beginning and End offer their thoughts on this issue:

When Moses sent his spies to scout the Promised Land so the children of Israel could enter it, the spies reported seeing beings there who were giants. The *Bible* makes it abundantly clear that there were races of people at that time who were so large, that regular sized men appeared as "grasshoppers" before them. . . . And we see that when the great servant of the Lord, Joshua, was sent to conquer the Promised Land that God commands him to kill all of the men, women and children of these giant races. They were to be wiped as if any remnant of their kind would post a great danger.[3]

Moving on, there is Numbers 13:31:

But the men that went up with him said, We be not able to go up against the people; for they are stronger than we. And they brought up an evil report of the land which they had searched unto the children of Israel, saying, The land, through which we have gone to search it, is a land that eateth up the inhabitants thereof; and all the people that we saw in it are men of a great stature. And there we saw the giants, the sons of Anak, which come of the giants: and we were in our own sight as grasshoppers, and so we were in their sight.[4]

And, with the most legendary and famous of all giants, there is Goliath, who was slain at the hands of David. Goliath is described in the pages of the Old Testament thus (1 Samuel 17:4–10 NIV): "His height was six cubits and a span. He had a bronze helmet on his head and wore a coat of scale armor of bronze weighing five thousand shekels on his legs he wore bronze greaves, and a bronze javelin was slung on his back. His spear shaft was like a weaver's rod, and its iron point weighed six hundred shekels. His shield bearer went ahead of him."

Another mysterious and enigmatic character who may very well have fallen into the category of Martian giants was Gilgamesh. And who, you might ask, was he? The answer is a sensational and eerie one, as you will now see. Our story all revolves around what is known as the *Epic of Gilgamesh;* it is an undeniably mighty tome that tells the story of the man (and king) of that name—a man who was not at all what he seemed to be. Without doubt, he was much more: He may not even have been wholly human—not fully, at least. Joshua J. Mark gets right to the heart of the story of Gilgamesh:

Gilgamesh is the semi-mythic King of Uruk in Mesopotamia best known from *The Epic of Gilgamesh* (written c. 2150–1400 BCE) the great Sumerian/Babylonian poetic work which pre-dates Homer's writing by 1500 years and, therefore, stands as the oldest piece of epic world literature. . . . Gilgamesh's father is said to have been the Priest-King Lugalbanda (who is featured in two Sumerian poems concerning his magical abilities which pre-date *Gilgamesh*) and his mother the goddess Ninsun (also known as Ninsumun, the Holy Mother and Great Queen).[5]

Giants in times long gone.
(Wikimedia Commons)

Not only that, Gilgamesh was also said to have been a demi-god—not a god, technically speaking, but the offspring of a human father and a supernatural mother. Zecharia Sitchin said of Gilgamesh, man or something else:

> Having been the son of the goddess Ninsun and the high priest of Uruk, Gilgamesh was considered not just a demi-god but "two thirds divine." This, he asserted, entitled him to avoid the death of a mortal. Yes, his mother told him—but to attain our longevity you have to go to our planet, Nibiru (where one year equals 3,600 Earth-years). So Gilgamesh journeyed from Sumer (now southern Iraq) to "The Landing Place" in the Cedar Mountains where the rocket-ships of the gods were lofted.[6]

Still on the matter of Sitchin, he may very well have been correct when it came to the matter of his theories concerning ancient astronauts coming to our world thousands of years ago. Quite possibly, though, Sitchin interpreted those same early extraterrestrials as the Anunnaki, when, in reality, he should have looked in a much closer direction: that of the planet Mars. Also, Sitchin's utterly absurd references to what he called "rocket-ships" should be noted. The idea that a trip, or multiple trips, from any faraway world would be successfully achieved by an armada of rocket ships is just plain ridiculous. To seriously muse on the possibility of traveling tens of millions of miles using brute force, rocket technology of the kind we employed to get NASA's *Apollo* missions to the Moon from 1969 to 1972 is, frankly, laughable. Despite this embarrassing, massive gaff on the part of Sitchin, there are other parts of the story of visitations to Earth from giant beings born on a faraway world that should be noted. For example, Sitchin had figured out that what he termed "The Landing Place" that so obsessed Gilgamesh was almost certainly Baalbek—a location we have already addressed and that is noted for its resident massive blocks of stones, which were built in a way that still baffles today.

Sitchin's thoughts on this were made very clear: "The great stone platform was indeed the first Landing Place of the Anunnaki gods on Earth, built by them before they established a proper spaceport. It was the only structure that had survived the Flood, and was used by Enki and Enlil as the post-Diluvial headquarters for the reconstruction of the devastated Earth."[7]

With that all said, let us now take a closer look at the life, and ultimately the death, of Gilgamesh. *The Epic of Gilgamesh* reveals that the king's grand plan, on becoming king, was to make the city of Uruk into a luxurious, sprawling city of epic proportions. He is said to have achieved the goal by using a huge number of slaves, all of whom overwhelmingly feared him. Day and night, they toiled, finally creating the legendary city that Gilgamesh so desired and demanded of his people. There is a very good reason as to why none of the slaves went against the ancient king—for a while, at least. It wasn't just due to the fact that Gilgamesh wielded incredible power and loyalty. It was also due to Gilgamesh's physical appearance. The tale of Gilgamesh states that while he was certainly human in appearance, there was something about Gilgamesh that really stood out from everyone else: He was incredibly tall. To be sure, it must have been an imposing and terrifying sight to have seen the King of Uruk looming—and no doubt booming—over one and all.

The time came, however, when even the slaves of Uruk decided that they wanted their freedom. Supposedly, they did so by communing with the Gods, who agreed to create a formidable foe, an equally powerful figure who could successfully take on the mighty strength of Gilgamesh. Notably, the legends say that this same foe—a huge, Bigfoot-type creature, no less—was fashioned from the saliva of humans. A nod in the direction of genetic manipulation and cloning, perhaps? Quite possibly. His name was Enkidu, a wild figure who was happy at home in woods and caves, and who loved animals. In the story of Gilgamesh, Enkidu crosses path with a prostitute named Shambat. Not surprisingly, she introduces Enkidu to the world of sex, as the epic tale makes abundantly clear:

> Shamhat loosened her undergarments, opened her legs and he took in her attractions. She did not pull away. She took wind of him, Spread open her garments, and he lay upon her. She did for him, the primitive man, as women do. His love-making he lavished upon her. For six days and seven nights Enkidu was aroused and poured himself into Shamhat. When he was sated with her charms, He set his face towards the open country of his cattle. The gazelles saw Enkidu and scattered, The cattle of open country kept away from his body. For Enkidu had stripped; his body was too clean. His legs, which used to keep pace with his cattle, were at a standstill. Enkidu had been diminished, he could not run as before. Yet he had acquired judgment, had become wiser. He turned back, he sat at the harlot's feet. The harlot was looking at his expression, And he listened attentively to what the harlot said. The harlot spoke to him, to Enkidu.[8]

Shamhat told Enkidu: "You have become wise Enkidu, you have become like a god. Why should you roam open country with wild beasts? Come, let me take you into Uruk the Sheepfold, To the pure house, the dwelling of Anu and Ishtar, Where Gilgamesh is perfect in strength, And is like a wild bull, more powerful than any of the people."[9]

It's not long at all before Enkidu—soon shunned by his animals in the woods, due to the fact that he had been exposed to sex and really enjoyed it—is told by Shamhat of the turmoil, slavery, and downright bad vibes that are presently going down in Uruk. So, as a result, Enkidu heads off to the city; his plan is to oust the much-feared king. As it transpired, however, things didn't exactly work out like that. Yes, there is a terrifying battle between the two giants—directly prompted by the fact that on his arrival at Uruk, Enkidu sees Gilgamesh trying to force a woman to have sex with him—but the outcome is not exactly what the people of Uruk are hoping for. Angered by what he sees before him, Enkidu charges at the giant king and a fierce and ferocious battle of the giants begins. The hand-to-hand combat goes on for hours, and although Enkidu is summarily defeated, by the end of it all Gilgamesh has developed a significant degree of respect for Enkidu. The outcome is that they become best buddies and go on a series of wild adventures, such as heading off to the Cedar Mountains—which, notably, Zecharia Sitchin believed was the Anunnaki's primary place of settlement.

On arriving, they encounter a character named Humbaba— also a giant-sized humanoid, and one who was noted for his incredible age—whose head they promptly and coldly remove from his body with a powerful blade. The two then encounter what was known as the Bull of Heaven, which is also referred to as Gugalanna, a deity whose origins were in the constellation of Taurus, better known as the bull. It, too, is killed. Then, there came a life-changing incident: When Gilgamesh refuses to have sex with Ishtar, a fertility goddess of Sumerian origins, the "gods" decide to take the life of Enkidu, by now Gilgamesh's best friend. Notably, Enkidu is killed by a virus. A bacteriological agent similar to those used on the battlefields of today's world on so many occasions? Possibly, yes.

There is a side effect to the death of Enkidu, one that plays on the mourning mind of Gilgamesh until the point of absolute obsession. It is—as you may have guessed—Gilgamesh's very own life. And, of course, his death, which he knows is inevitable, even if he's not precisely sure when and where he might take his final breaths. Enkidu's death, and Gilgamesh's own fears combined, lead Gilgamesh on a trek to seek out the secret of immortality. It should be stressed that Gilgamesh had very little to complain about; after all, he is said to have ruled over the people of Uruk for significantly more than a century. That fact alone suggests that the claims that Gilgamesh was a demi-god—or, perhaps, someone who was half-Martian and half-human—were the truth.

In his quest to keep the reaper firmly at bay—forever, possibly—Gilgamesh sought out a man named Utnapishtim, who, in the *Epic of Gilgamesh,* is the mirror image of the Old Testament's Noah. In the *Epic of Gilgamesh,* Enki, of the Anunnaki, shares with the ancient king something terrifying: The world will soon be overwhelmed by just about the worst deluge possible. Lands—in fact, entire nations—will be swallowed up by the massive waves that will soon lay waste to most of human civilization. And, so, a panicky Utnapishtim embarks on the construction of a huge boat. It's known as the Preserver of Life—a most appropriate title, to be sure. Also, the "seed" of every animal is carefully stored on board the ship. It's interesting to note that whereas the Old Testament tells of the animals being taken on board Noah's Ark two-by-two, it's the "seed" of all the animals that Utnapishtim is tasked with preserving. That very word—*seed*—strongly suggests that it was the DNA, the cells of the animals that were preserved, no doubt to allow for future cloning—and, effectively, "resurrecting" the animals whose essence is safely aboard the Preserver of Life. For taking on the huge role of dutifully looking after the seeds of the animals, as well as the people chosen to survive the flood, the gods bestow upon Utnapishtim and his wife the much-sought-after secrets of immortality. So, now, we see why Gilgamesh was so desperate to meet with Utnapishtim: He wanted those secrets of immortality for himself.

Long after the flood waters and chaos receded, Gilgamesh finally finds Utnapishtim, who, at the time, is living on an isolated island with his (Utnapishtim's) wife. It's thanks to a man named Urshanabi, who knows the secrets of how to negotiate the ocean waves and leads Gilgamesh to Utnapishtim. Despite his best efforts, Gilgamesh is shot down by Utnapishtim. While Utnapishtim fully understands Gilgamesh's fears of death, Utnapishtim tells the king that there is very little he can do to help him. There is, however, a glimmer of light at the end of the tunnel. It comes from none other than Utnapishtim's wife.

Gilgamesh is told that, to a degree, she can help him on his quest for youth and a massive lifespan. The answer to the riddle comes by ingesting the contents of a mysterious plant that can only be found on the bed of the oceans. Supposedly, Gilgamesh finally—to his eternal joy—locates the much-needed plant. He is just about to partake in the plant when a monstrous, coiling, sea-serpent-type beast looms into view and suddenly snatches the plant away. Gilgamesh, psychologically crushed, realizes that immortality—or something very close to it—is now out of his hands. With the inevitably of death now on his mind, Gilgamesh decides to cease his quest for eternal life and heads back to Uruk, where he lives out the rest of his life. That he ruled over the city and its people for 130 years—and was probably at least in his twenties when he took to the throne—suggests he was indeed not entirely human and had a life much longer than we can ever hope to attain.

Many of you reading these words might roll your eyes and suggest that the *Epic of Gilgamesh* is nothing but a fanciful, entertaining, old tale. Perhaps that is exactly what it was. On the other hand, though, there is available, fragmentary evidence in favor of the possibility that at least certain parts of the story can be validated. For example, in the early part of 2003, there was a major development in the goal to further understand the truth of the *Epic of Gilgamesh*. It was all very much thanks to Jorg Fassbinder, of Germany's Bavarian Department of Historical Monuments. He and a team of crack archaeologists headed off to what was once ancient Uruk, secured permission to excavate and dig, and came up with something sensational. On April 29, 2003, the BBC ran an article with the eye-catching title of "Gilgamesh Tomb Believed Found." It began: "Archaeologists in Iraq believe they may have found the lost tomb of King Gilgamesh—the subject of the oldest 'book' in history. . . . Now, a German-led expedition has discovered what is thought to be the entire city of Uruk—including, where the Euphrates once flowed, the last resting place of its famous King."[10]

When asked for his thoughts on the potentially extremely important development, Fassbinder told the media:

> I don't want to say definitely it was the grave of King Gilgamesh, but it looks very similar to that described in the epic. We found just outside the city an area in the middle of the former Euphrates river; the remains of such a building which could be interpreted as a burial. By differences in magnetization in the soil, you can look into the ground. The difference between mudbricks and sediments in the Euphrates River gives a very detailed structure. We covered more than 100 hectares. We have found garden structures and field structures as described in the epic, and we found Babylonian houses.[11]

Clearly, the discovery was an amazing one. And, while the debate still goes on as to whether the finding really was the tomb of Gilgamesh, there is a good, solid, likelihood that this is precisely what it is claimed to be. If, one day, we should we find the remains of Gilgamesh—maybe even a complete skeleton—and it proves to be somewhere in the region of sixteen feet in length, then this will be a development of totally unprecedented proportions, one that will surely add even more controversy to matters relative to human origins and ancient giants—from here and there, wherever "there" may be. Mars, perhaps.

In similar territory, in late 2015 there was yet another significant development in the story, as Ted Mills, at *Open Culture*, demonstrated. He said, "One of the oldest narratives in the world got a surprise update last month when the Sulaymaniyah Museum in the Kurdistan region of Iraq announced that it had discovered twenty new lines of the Babylonian-Era poem of gods, mortals, and monsters."[12]

We just might be on the verge of uncovering yet further information on the giant king who, incredibly, may have been part-Martian. Such an admittedly controversial theory is not at all impossible. Lest we forget: During the CIA's remote-viewing session of May 1984, the viewer described seeing on Mars figures that "appear thin and tall." They are also referred to as being "very tall" and "very large."[13] And, on the matter of Ingo Swann's remote-viewing operations in the latter part of the 1970s, the man was able to view huge humanoids of heights up to no less than ten feet. And, Swann was able to deduce that those same giants were Martians.

CHAPTER 23

"They Created
a Permanent
Space Base on Mars"

||||||||||||||||||||||||||||||||||||||||

N ow, it's time to address the ultimate, most famous giants of all: the Anunnaki. In my 2015 book, *Bloodline of the Gods,* I said that at some point in the distant past:

A mighty race of legendary beings came to our planet from the heavens above—and, during their time here, brought some form of stability, and even society, to what were extremely primitive human tribes. That was not, however, their original agenda, as we shall soon see. As for the physical descriptions of the Anunnaki, theories vary from giants of eight to ten feet tall to bipedal reptiles, and with some researchers concluding they may well have been both—the power of shape-shifting allowing them to manifest in various guises. This particular scenario, referred to above, of the Anunnaki performing a teaching role to the people of Earth was a staple part of Sumerian beliefs and lore, and remains an important piece of historical record. But, who, exactly, were the Anunnaki?[1]

To answer that question we have to turn our attentions to Lawrence Gardner, whose books included *Genesis of the Grail Kings: The Explosive Story of Genetic Cloning and the Ancient Bloodline of Jesus*. It's a book that delves deep into the world of the Anunnaki. Gardner said of the Anunnaki:

They were patrons and founders; they were teachers and justices; they were technologists and kingmakers. They were jointly and severally venerated as archons and masters, but there were certainly not idols of religious worship as the ritualistic gods of subsequent cultures became. In fact, the word which was eventually translated to become "worship" was avod, which meant quite simply, "work". The Anunnaki presence may baffle historians, their language may confuse linguists and their advanced techniques may bewilder scientists, but to dismiss them is foolish. The Sumerians have themselves told us precisely who the Anunnaki were, and neither history nor science can prove otherwise.[2]

While Gardner certainly flew the flag of the Anunnaki, in terms of the theory as to who they were and the nature of their agenda on Earth, there's no doubt at all that the primary figure who brought the world of the Anunnaki to millions of people was Zecharia Sitchin. He wrote extensively on the subject of these giant, ancient aliens, including *When Time Began, The Wars of Gods and Men,* and *The Cosmic Code.* Sitchin grew up in the Azerbaijan Soviet Socialist Republic, studied economics at the University of London, England, and immigrated to the United States in the early 1950s. His fascination for the tales of the Anunnaki—one might even say his outright obsession—led him down a controversial and amazing pathway. As time and his research progressed, Sitchin came to the realization, in his mind, at least, that the Anunnaki were not supernatural gods. Quite the opposite, in fact. In Sitchin's mind, the Anunnaki were nothing less than a mighty race of extraterrestrials that came to the Earth in the extremely distant past.

The agenda of this ancient race was (1) to genetically alter—upgrade, some might say—early, primitive humans and use them as a slave race; and (2) to mine the Earth of its gold, for a controversial reason that will soon become apparent. Sitchin also concluded that one particular aspect, perhaps more than any other, led early humans to view the Anunnaki as nothing less than gods: It was Sitchin's conclusions that the Anunnaki had cracked the secret of immortality—or, rather, of near-immortality. Yes, the Anunnaki could die. And they did die. Their lifespans, however, suggested Sitchin, were practically mind-boggling from our perspectives as members of the human race. He made the claim that the Anunnaki lived for hundreds of thousands of years; a far cry from our measly eighty or ninety years. And there is one more vital component to the story. It's the tale of the home-world of the Anunnaki, which Sitchin claimed was called Nibiru.

According to Sitchin, Nibiru is a massive world that orbits our very own sun, but that is only visible when its orbit brings it close to the Earth—perilously close, too—approximately every 3,600 years. Sitchin's research led him to believe (and, it must be stressed, much of what Sitchin concluded *was* belief-driven) that the Anunnaki arrived on Earth hundreds of thousands of years ago, landing somewhere in what today is known as the Persian Gulf. Now, it's time to address the matter of the Annuaki obsession with gold.

Sitchin—as well as the late Jim Marrs, whose conclusions dovetailed closely with those of Sitchin—suspected that part of the reason why the Anunnaki needed gold, and needed a great deal of it, was because of its life-extending properties. *Token Rock* says of this strange angle:

Throughout history, alchemists have sought the elusive Philosopher's Stone, the secret White Powder Gold which would become quite literally a vessel of the "light of life." This secret material was reported to bestow powers of immortality in addition to incredible supernatural powers to those who consumed it. Certain famous mystics, magicians and alchemists of history like Enoch, Thoth and Hermes Trismigestus are known to have perfected the sacred art of creating The Philosophers Stone and their use of the material explains the many legendary supernatural powers ascribed to them.[3]

Whether you buy into Sitchin's controversial theories or you don't, it is a fact that accounts of mysterious substances, with the ability to extend life to massive degrees, absolutely abound in old world tales. The Philosopher's Stone, Manna from Heaven, Amrita and Ambrosia are just some of the mysterious cocktails that promised one would never have to die. John 6:50–51 (NSB) says the following of Manna: "This is the bread which comes down out of heaven, so that one may eat of it and not die. 'I am the living bread that came down out of heaven; if anyone eats of this bread, he will live forever; and the bread also which I will give for the life of the world is My flesh.'"

Now, let's have a look at what Sitchin concluded concerning a second reason as to why gold was so important to the Anunnaki. Although Nibiru was millions upon millions of miles away from the Earth, and it was much larger than our world, it did have something in common with the Earth: In just the same way that we, today, are concerned by the thinning of the planet's ozone layer, so did the Anunnaki, whose world was said to have been in a far worse state then than ours is right now. And, how did the Anunnaki seal those growing holes in its ozone layer? They used gold. Sounds strange, right? Definitely. But read on. The story isn't so weird, after all.

The US Environmental Protection Agency has significant concerns about the decaying ozone layer of our planet:

Ozone in the stratosphere, a layer of the atmosphere nine to 31 miles above the Earth, serves as a protective shield, filtering out harmful sun rays, including a type of sunlight called ultraviolet B. Exposure to ultraviolet B has been linked to development of cataracts (eye damage) and skin cancer. Scientists have found "holes" in the ozone layer high above the Earth. The 1990 Clean Air Act has provisions for fixing the holes, but repairs will take a long time.[4]

The late Lloyd Pye, author of *The Starchild Skull: Genetic Enigma or Human-Alien Hybrid?*, was convinced that gold—or, rather, gold flakes—in massive amounts had driven the Anunnaki to travel to Earth and plunder its gold. Plugging the holes in the ozone layer of a planet with millions of tons of gold flakes? To be sure, it sounds like wild science-fiction. Incredibly, though, it's actually not.

In December 2009, the US House Select Committee on Energy Independence and Global Warming held a debate on matters relative to such pressing issues as global warming and the degrading of the ozone layer. Dr. John P. Holdren suggested flooding the higher levels of the atmosphere with what were termed "pollutants."[5] In effect, plugging the holes. There was nothing new to this, however. Back in the 1970s, Dr. Edward Teller, the "father of the H-bomb," came up with just such an idea. His idea—almost identical to the theory Sitchin promoted with regard to the agenda of the Anunnaki—was to swamp the upper atmosphere with heavy metal-based particles. This would, hopefully, ensure that the ozone layer would stabilize—and even be repaired. The result: the damage to the ozone layer would slow down and eventually cease.

As for who decided that the Earth should become the "feeding ground" of the Anunnaki it was, said Sitchin, Enlil and Enki—who were said to have been the sons of the Anunnaki overlord, Anu. When Nibiru was at its closest to Earth, a veritable armada of spacecraft headed to Earth. On arriving, in what was once called Mesopotamia (today, it's the Tigris–Euphrates River system that covers portions of Turkey, Iraq, Kuwait, Syria, and Iran), they very much liked what they saw and found: an abundance of gold and a race of beings (us) that could be genetically "altered" and that would ultimately become the slave race of the Anunnaki—toiling away in massive, almost unending gold mines all across our world. According to Sitchin, though, the vast majority of the work was undertaken in the gold mines of Africa.

Those doubtful and dubious of such an undeniably fantastic scenario, and who may demand evidence of such mining, should note that back in the 1970s the Anglo-American Corporation, a mining consortium based in South Africa, uncovered evidence of ancient mining activity in the country, all of which was estimated to have occurred at least 100,000 years ago—by whom is a question that remains notably unanswered. It's *also* notable that in 1997 Ian Smith, the last prime minister of Rhodesia, which is now Zimbabwe, called South Africa "one of the most richly mineralized parts of our world."[6] Indeed, in 2002 alone, the country produced no less than fifteen percent of the world's entire gold output.

Sitchin's grand theory grew and grew. It involved revolution on the part of the slave workers, and of Nibiru coming perilously close to the Earth—and, as a result, causing massive, worldwide disaster to our planet due to the effects of Nibiru's gravitational effects. Those ancient tales of mega-sized floods—whether localized or worldwide—were driven by what happened when the veil between the two worlds quickly shrunk, Sitchin believed. The Anunnaki, realizing that absolute disaster was just around the corner, were able to save not just many of themselves, but of the human race too—hence the story of Noah and the flood in the Old Testament, and of the earlier account of Utnapishtim, which just happens to appear in *The Epic of Gilgamesh*.

The destruction of Sodom and Gomorrah by nuclear weapons, and the "fact" that the Anunnaki began to turn on each other as their long-running reign began to spiral into chaos made matters even worse, Sitchin told his readers. The result: This race of ancient, near-immortal giants finally left the Earth for Nibiru, never to return. As I noted earlier in this chapter, much of what Sitchin had to say was rank speculation and not a lot more. His overall scenario was a theory—an intertwining of countless ancient legends designed to show how and why ancient extraterrestrials came to the Earth, radically changed the human race, and plundered our planet for their own means. And, it should be stressed, not a single piece of evidence can be found to show that Nibiru ever existed. Also, as you will have deduced, there are distinct issues concerning how much of all this covered a period of time of a few thousand years, tens of thousands of years, or even of hundreds of thousands of years. The picture, then, is admittedly confusing and driven by speculation.

This has all led to a fascinating, very different, scenario—one that sits comfortably in the story that this book tells: that the Anunnaki were *not* from the wholly 100-percent-elusive planet known as Nibiru, but from our old friend Mars. In that sense, Sitchin may have been broadly *correct* in his theories—of visitations to the Earth by extraterrestrials hundreds of thousands of years ago and of "alterations" to primitive humans—but spectacularly *wrong* in terms of the world from which the Anunnaki *really* came. Or, the story might be even more complex than that: Perhaps Nibiru did exist (and, maybe, somewhere, it still does), but the activities of the Anunnaki were not just limited to the Earth. They may have established outposts on Mars, too. It should be noted that Sitchin actually suggested that this was a part of the larger, overall story.

"In their texts the Sumerians wrote that the Anunnaki traveled to Earth in groups of fifty. The first team," Sitchin explained, ". . . splashed down in the waters of the Persian Gulf [and] waded ashore." Sitchin had more to say:

"In time, 600 Anunnaki were deployed on Earth and another 300 operated shuttlecraft between Earth and Mars. Yes, Mars!" He continued: "The Anunnaki, I have concluded in my books, used Mars not just for a quick stopover; they created a permanent space base on Mars, complete with structures and roads. In *Genesis Revisited* I reproduced numerous photographs taken by NASA's Mariner-9 in 1972 and Viking-1 Orbiter in 1976 that clearly showed a variety of artificial structures there. Some of them were in the Cydonia area with its famed Face."[7]

So, we have an even more complex possibility: that, while engaged in its massive activities on Earth, the Anunnaki established facilities on the Red Planet—a world that was possibly *already* teeming with its very own entities: the Martians. So, we are talking about three worlds, all harboring life to varying degrees. In the next chapter, we'll see that there is evidence, secured by a skilled remote viewer, to support this theory that the story of the Anunnaki is inextricably linked to not just the Earth and Nibiru, but to Mars, too.

The final words—on this issue, at least—go to Chris H. Hardy. He has a doctorate in ethno-psychology and is a cognitive scientist and former researcher at Princeton's Psychophysical Research Laboratories. Hardy concludes, "The nuclear bombing of five cities of the Jordan plain, including Sodom and Gomorrah" directly caused "the destruction of the Sumerian civilization and the Anunnakis' own civilization on Earth, including their space port in the Sinai."[8] Hardy has also explored the possibility of the Anunnaki having once had "bases" on Mars, which adds an entirely new dimension to the situation concerning the flattened cities.

As Hardy's 2014 book, *DNA of the Gods*, notes:

[T]he Anunnaki came to Earth from the planet Nibiru seeking gold to repair their ozone layer. Using genetic engineering, they created modern humanity to do their mining work and installed themselves as our kings and our gods. Anunnaki god Enki had a fatherly relationship with the first two humans. Then Enlil, Enki's brother, took over as Commander of Earth, instating a sole-god theocracy and a war against the clan of Enki and humanity for spoiling the Anunnaki bloodlines through interbreeding. This shift imposed a blackout not only of the very human nature of the Anunnaki "gods" but also of humanity's own ancient past on Earth.[9]

Now, we come to another remote viewer, who in 2019 was able to put together many of the threads that run throughout this incredible story.

CHAPTER 24

"There Are Thunderous Vibrations that Shake Mars"

|||||||||||||||||||||||||||||||||||||||

Dr. Kimberly Rackley, MscD, is a skilled psychic intuitive and someone who has an extraordinary story to tell that revolves around the final, tumultuous days of the Martian civilization. A noted remote viewer—not at all unlike those hired by the CIA in the 1970s—Rackley is someone who I have consulted on a number of occasions and on a variety of subjects. In October 2019, Rackley agreed to do something for me that was not at all dissimilar to what the CIA did back in 1984: namely, to supernaturally surf both history and a faraway world—Mars, of course—and to try to determine what, exactly, happened to Mars and the beings that once lived there. Whereas the CIA's remote-viewing session concerning Mars revolved around a specific time frame (one million years BC), Rackley's did not. Rather, her approach was to address the lead-up to the disaster and how and why it came to be. We'll begin with what Rackley saw of the terrain on Mars before that planet-wide disaster changed everything—and forever.[1]

Rackley told me that she focused her attention on one specific portion of the Martian world; but, importantly, a portion that was representative of Mars as a whole. The planet, in the past, was not that different to ours, Rackley said, as she targeted Mars's past: "Greenery" dominated the landscape, Rackley said, adding and stressing that "the growth is very lush." And mostly near areas dominated by water. She was able to "see" clean, clear rivers and lakes, "rocky areas, cliffs, and mesas. The water was very clear; it was like sapphire blue clear. I wasn't afraid of anything that might be in the water; it was totally clear." By all accounts, Mars was a veritable paradise. Rackley expanded:

> There were oasis' [sic] which looked very normal, except that the trees looked like palm trees, but different, too. It was a bunch of little sites of trees, rather than big ones. Not forests. There were fruits on the trees, and obviously I don't know what kind of fruit. They were orangey, like an orange-red beet color. And there were flowers that looked like orchids. They may not have been orchids, obviously, but that's the closest thing I could relate to: orchid-type flowers.

Now, we get to the incredible matter of one particular creature that lived on Mars's surface. Rackley reveals what she encountered in her altered state:

> I saw crab- or spider-like creatures, rather large with eight arms around the shell of the

Kimberly Rackley, who remote-viewed the final, fatal days of Mars. (Kimberly Rackley) body. Sensors are at the end of these arms, but I see no claws or fingers. The shell appears as a shell within a shell, with red and various browns in color. While the body is large, perhaps the size of a cow, the head seems to be very out of proportion. The head is part of the shell, as I see no neck. There are wide, large eyes and slits that I presume are the noses. The mouth appears as a slit as well, and at the time of this viewing I could not detect teeth. These crab creatures eat bacteria found in the soil and within the water.

It's vital and appropriate that I interject here: When Rackley described the physical appearances of the creatures she saw, it instantly reminded me of the so-called face-hugger-type animal captured on camera by NASA—which are content to conclude that simulacra was the cause of the sensational story. After speaking with Rackley, who had no previous involvement in investigating matters of a Martian kind, I showed her the face-hugger photo, which undeniably shocked her to her very core. In response, and after expressing a high degree of amazement, but also a deep feeling of satisfaction, she said: "That tells me what I was connected to was right on." She had more fragments of this viewing to share with me, too: "I got the idea the crab things were more animal—I mean, they had intelligence, obviously; every animal does—but, they weren't like us. They had no technology, no buildings."

Now, it's time to turn our attentions to, incredibly, a *second* life-form on Mars.

Rackley said:

I then witness what I can only describe as ant-like creatures. They walk on two legs and they have three fingers and four toes. I see two antennas. The creature is perhaps five-feet tall and reddish-black in color. Their eyes are like large eggs; oval-shaped. The brain is an egg-shaped organ, as well. They seem to be androgynous as I can see nothing that indicates gender. I see shafts or tunnels underground associated with these creatures. Within some of the shafts, there are pods that are thin and elongated. I see no weapons. I see little crevices where two to four creatures gather; I feel it is a family unit. This underground system is vast. Mostly it appears dark and I can see in some places of the walls there are shimmering elements. Perhaps ores and the like.

When I addressed the matter of the ant-like creatures, Rackley said that she felt these were "the real Martians." She opened up further on the nature of the realm of the ant-like creatures: "It reminded me of a giant ant farm, too; the kind we see [on Earth]. But more refined. Those little pods that were scattered in various places: I don't know if they were for traveling along underground tunnels, or if the technology was used to make the underground shafts. I got a very strong sense the ant creatures were the original inhabitants of Mars."

Now, we enter even more controversial territory. It takes us back to the world of the Anunnaki and how their civilization came into play in this ultimately tragic saga. Notably, one of the first things that Rackley saw was the image of one of the Anunnaki. The male entity she encountered was "very tall, like really sinewy-looking. Bulky arms, muscular. I didn't see any hair, but I could see on his chin that there might have been something like that, but I'm not sure. The eyes were large. Like a wide oval shape. Maybe ten or eleven feet tall. He wore a strange outfit; an outfit almost built into him." The final word that Rackley used to describe the Anunnakian in her midst was "imposing."

In a further sitting, Rackley concentrated on a portion of Mars's landscape. It was dominated by what were clearly the ruins of massive structures; the most astounding one being a huge pyramid, albeit a noticeably severely damaged pyramid. The specific location on Mars was unknown to her. Rackley was not sure that what she was seeing now—namely, the ruins—occurred in the same Martian time frame in which she saw the ant-like creatures, the Anunnaki giant, and the crab-creatures going about their activities. The only thing she could be sure of was that everything was intertwined in terms of Mars's mystery-shrouded history. In all probability, it *was* two or more time frames. After all, Rackley's initial quest for answers revealed Mars to be a beautiful, lush world. That she later had images in her mind of certain areas of Mars that were completely laid to waste, suggests strongly that something had gone very wrong with the Red Planet—*very* wrong. We'll soon find out what that "something" was.

According to what Rackley learned during her remote-viewing session, the Anunnaki used Mars as a kind of outpost on their journeys to our Earth, and had done so for thousands of years. As a result, they had created permanent facilities on Mars: pyramids, for the most part. Rackley added: "I didn't see any specific [planetary] coordinates; I just saw how it was set up: there was a lake, and a settlement, and the ruins were to the left bank of the lake."

Of some of the extensive construction that had occurred on Mars, Rackley provided these words:

The structures I saw were a combination of ships—where the Anunnaki would stay inside—and stone buildings to protect them from the Martian atmosphere. And the entire area has a huge shield. Or, what looks like a massive bubble all around it. The structures have stabilized air within them. I see they mainly wear helmets outside of the ships and structures. Perhaps the Mars air was not easy to breathe, for the Anunnaki, even when Mars was still a healthy world and wasn't dead.

There was more to come. She continued:

I do not see structures of any kind further back than 400,000 years. The pyramid structure I see serves as a landing beacon and a generator of power. The pyramid housed crystals that generated energy. The pyramid was made by a machine that slices stone with great accuracy that was developed by the Annunaki. This same slicing process was responsible for the great pyramids in ancient Egypt. As well as other pyramids in other lands. A landing pad was behind the pyramid; it may be a mesa or a smoothed-out, raised landmass, although with large chunks now missing from it. I was shown that Baalbek served the same purpose.

As we have already seen, Baalbek and the Anunnaki were already intertwined long before Rackley came onboard. Interestingly, Rackley said that despite what many have written about an apparent, or assumed, huge presence of the Anunnaki living on our world and manipulating us on a massive scale, that was not actually the case. Yes, she said, their presence had an incredibly influential effect on our early legends, myths, cultures, and religious beliefs, but they were not overly huge in numbers. In fact, at the very most, the numbers of personnel that the Anunnaki had on Mars and the Earth at any given time only amounted to around several hundred at most, said Rackley.

She added: "The ant-creatures seemed to live alongside the Anunnaki; they mainly kept to themselves—and the Anunnaki did the same. I didn't see any interaction between them. The ant people meant nothing to the Anunnaki. And, it was the same the other way around. The Anunnaki used Mars like we would an off-shore military base, between Nibiru and Earth and that was it."

Now, we get to the finale: the shuddering, terrifying end of civilization on Mars and the Anunnaki's intervention to try and salvage some degree of the Martian culture and the planet's "people." According to what Rackley saw, a long, long time ago Nibiru's orbit provoked massive, chaotic activity in what we, today, call the Asteroid Belt. On the matter of asteroids, NASA provide this:

Asteroids, sometimes called minor planets, are rocky, airless remnants left over from the early formation of our solar system about 4.6 billion years ago. The current known asteroid count is:

823,318. Most of this ancient space rubble can be found orbiting the Sun between Mars and Jupiter within the main asteroid belt. Asteroids range in size from Vesta—the largest at about 329 miles (530 kilometers) in diameter—to bodies that are less than 33 feet (10 meters) across. The total mass of all the asteroids combined is less than that of Earth's Moon.[2]

Rackley did not learn that Nibiru's orbit utterly shattered and destroyed a now-unknown world and, in the process, created the Asteroid Belt. Quite the opposite, in fact. Her remote-viewing session revealed that the Asteroid Belt was already in existence— and had been for a very, *very* long time. The gravitational effects of Nibiru, however—which apparently had a huge orbit and that only crossed paths with some of the inner planets of our solar system occasionally—caused such near-unimaginable chaos and destruction that a huge number of already-existing asteroids were violently wrenched out of their orbits in the Asteroid Belt. In almost-sling-shot style, and like an unstoppable, mighty, stone armada of death, the massive rocks were driven toward Mars by Nibiru's gravitational activity. It was something that not even the Anunnaki were able to bring to a halt. In a disastrous fashion, the gigantic rocks got closer and closer and in relentless, head-long fashion. The clock was ticking. And ticking. That is, until those now "free" asteroids finally careered through the Martian atmosphere, slamming down on Mars and causing irreversible, planet-wide chaos and destruction.

Seas were boiled dry. As were lakes and rivers. The sky was filled with choking smoke. Millions of tons of dust blotted out the light of the Sun. Massive quakes shook the entire planet. Those glorious pyramids were, at best, damaged beyond repair. At worst, they were reduced to pulverized rock, barely recognizable for what they once were. The secrets of the purposes of the pyramids were gone. In short time, Mars went from being a beautiful world to a veritable dead zone. An untold number of Martian lifeforms were killed. As for Mars's flora: all but gone, too.

There was, however, says Rackley, just enough time to save at least some of Mars's beings, and its cultures. Rackley says: "I saw that at some point when the Anunnaki went to Earth, and left Mars, they took some of the ant-people with them. The Anunnaki knew the atmosphere of Mars was going to decline quickly now that the asteroids were starting to hit, and they agreed to take them to Earth when the asteroids were coming down." Even though the two—the Anunnaki and the ant creatures—didn't interact before, this was different: It was so they wouldn't eventually be annihilated. The time came when the atmosphere of Mars was beyond repair. It would continue to dissipate for some time. The Anunnaki left in their ships and either stayed in orbit or went straight to Earth. Those were the only choices for everyone.

And there is one more issue that I asked Rackley to dig her teeth into, so to speak: that of the Twin Peaks and the Sphinx, as they have controversially become known. In a later, October 2019 remote-viewing operation, she shared with me her findings while surfing both past and present in skilled fashion. Getting right to the heart of the mystery, Rackley began as follows:

> I began by concentrating on the Twin Peaks. As I did, I was almost immediately able to determine one important thing: My first impression of the Peaks, themselves, is that they are not entirely natural. I see a bright blue color beyond them. From what I could view, this was definitely indicating a water source to me. At this source of water, I see what I call a spaceship taking on water through some kind of tube. I get a feeling that from time to time there are still occasional visits to this area by what I believe are the Anunnaki. Maybe just two or three of them; that's all. It's as if they are checking in on the location, now and again, with a small team.
>
> Under my feet, I feel that deeper into the ground there is much more than can be seen on the surface of Mars. As I focused my

attention under the sand and rock, I see a maze of pathways, chambers, and other structures. Some of these paths lead directly to the peaks: underground. At the same time all of this is going on, I'm hearing a voice say that "Others from where you came have information that is kept hidden." I see a NASA rover has been very close to this area and I see that data is being stored and transmitted back to Earth. This tells me NASA is keeping something hidden about this place.

The data on the Twin Peaks secured during Rackley's remote-viewing session pinpointed on what could be seen, and learned, specifically today—in our time. During the process, however, her mind was suddenly flung into the distant past. She suspected it was to a point around 150,000 years ago. It was in this mindset that Rackley was able to get a glimpse of how the Twin Peaks looked way back then. What she saw amazed her, and that's putting things mildly. What Rackley encountered right in front of her were the Twin Peaks when they were (no pun intended) at their absolute peaks, standing proudly on the Martian landscape, and clearly fashioned by an advanced intelligence. She explained that the pair looked precisely like what one would expect to see if one traveled to Giza, Egypt, and secured a few photographs of our own legendary pyramids. And, there were humanoid beings around the Twin Peaks, as she revealed:

I see tall human beings; they are very striking in their features. They have prominent cheeks, nose, and foreheads. Their skin varies from olive to white. Some of them are dressed in attire that is much like that worn by the ancient Romans and Egyptians. Or, kind of like a combination of the two. Upon their wrist are bands with a star encased by a circle. And, from the star little rays are shooting outward; I don't know what their purpose is. Some are wearing helmets that have what look like wings on each side.

Another glimpse of the Anunnaki on Mars? Or of yet *another* form of indigenous Martian? We are unable to answer those questions. Now, the story gets disturbing.

We have already seen how Rackley visualized much of Mars destroyed beyond repair, as well as the ultimately inevitable spiraling of the planet into a pale shadow of what it once was. While the overwhelming amount of destruction was caused by those all-powerful asteroids, in the last days of Mars, said Rackley, pockets of the surviving Anunnaki who had not already fled Mars for the Earth turned upon each other, almost certainly with nuclear weapons.

She added: "My attention is drawn to the sky, specifically in the area between Mars and Earth. I am not comfortable with what I see and I become a little anxious. There are ships in a battle against each other. The ships are different in size, but of the same people. They are warring against each other. Heat engulfs me. White-hot heat. There are thunderous vibrations that shake Mars and I see horrible, huge stinging vapors in the sky."

The end really was nigh.

Now, it was time for Rackley to see what she was able to see with regard to that Martian Sphinx today. The data makes for fascinating reading:

The sphinx reached deep into the ground and it still does. I am being shown a being of great stature—a giant—with wealth and knowledge in connection to the Sphinx. The sphinx represents this being as well as a place of origin: the Anunnaki's original home-world. There is a large chamber here. Symbols are imposed upon the walls in some areas. I see the walls were once lit up or were activated but the source of power is no longer present.

In just a few words, Rackley gets to the poignant truth: "Burnt out or removed I cannot tell. There is only a faint memory vibration left. Everything else is gone."

One final thing: There is the matter of those ant-people that Rackley said were the "original inhabitants of Mars." When she described their physical appearances to me, it instantly made me think of the ant-people who are an integral part of the legend, history, and mythology of the Hopi Indians of northern Arizona. Hopi lore tells of how, in times long gone, the Earth underwent a series of terrible cataclysms (or cycles) that manifested in the forms of Ice Ages, massive floods, and possibly even calamitous polar tilts. So the story goes, the Hopi might well have been exterminated when the world was in massive flux, were it not for assistance given to them by a race of diminutive, mysterious, humanoid beings. They were said to live deep underground, in certain parts of the American southwest. They were, of course, the Ant People.

According to the legends, for the most part the Ant People kept away from the human race—which was largely oblivious to the existence of the strange world deep below them. On occasion, however, the Ant People would surface and, also on occasion, offer help to people. Thus, bit by bit, the Ant People became known and the legends of these entities duly began to develop.

It's important to note that they weren't called the Ant People just because they lived underground in extensive, deep, winding tunnels. It was also because, despite being humanoid in stature, they somewhat physically resembled ants, particularly so in terms of their faces and their spindly limbs. Of course, one doesn't have to be a genius to know that some of today's so-called "Grays" of alien abduction lore superficially look insect- or ant-like.

Are today's Grays the descendants of those Ant People saved by the Anunnaki millennia ago, taken from Mars in its final days, and brought to Earth, as Kimberly Rackley described? And another question: What if at least some of those Hopi legends (dominated by tales of ant-like entities) of polar tilts, huge deluges, and massive cataclysms were not memories of what occurred on our world in the distant past, but on Mars, in an era almost exclusively forgotten and lost?

Now, we come to the strangest part of this particular story of remote viewing and the end of Mars. The process of having Rackley perform the remote-viewing experiments, my follow-up questions, and our many emails, smart-phone calls, and texts took up a period of time close to a week. It was during that very same period that both of us noticed something weird: We found ourselves dominated by the numbers 11:11. I would see them when I randomly looked at the time on my laptop and on my cell. I even saw it flashing on my microwave after a power outage late one night. One evening, a friend called asking about the Men in Black (MIB). Two days earlier, she had an encounter with a MIB outside of her home. She just happened to phone me at 11:11 p.m. to share the story. Rackley had her own equivalents—over and over again. It was, to say the least, a crazy time. So, to where is all of this leading? Well, the 11:11 phenomenon is one that many are aware of, but that others are wholly unaware of. I have brought the matter up with people many times. Sometimes it results in knowing nods. On other occasions, though, it provokes nothing but blank faces. Let's see what those in the know say about the phenomenon.

The website Dimension 11:11 states:

What does 11:11 mean? What is the significance of seeing repeating number patterns such as 11 11? There are many people who feel that there is nothing special or out of the ordinary about seeing the time 11:11 on a digital clock or watch. But for others, seeing these numbers frequently showing up in their lives gives them a peculiar or surreal feeling and it often becomes a meaningful experience to them.[3]

Then, there are the words from the Power of Positivity website:

What does it mean if you keep seeing 11:11? Our Spirit Guides, angels, or higher selves like to speak to us through various methods, such as playing a recurring song on the radio that may have special significance, answering a prayer, flipping to a certain page in a book we're reading, or even directing our attention to repeating numbers on a clock or sign, such as 11:11. At first, this occurrence might seem like a silly coincidence, but by looking further into it, you will find that it has a powerful spiritual message hidden within.[4]

It was all of this strange activity that led Rackley to ponder on an incredible possibility: that her remote viewing of Mars may have caught the attention of the Martians or the Anunnaki themselves! Bizarre? Yes! But, I can absolutely attest that that particular week was filled to the brim with synchronistic weirdness of the 11:11 type. Take it or leave it, that's exactly how it all went down, for both of us. I wondered if—in just the way that Mr. Axelrod warned Ingo Swann that by remote viewing the Moon the Martians were now able to turn the tables and focus on Swann himself—Rackley and I were now ripe for Martian scrutiny of the secret kind. Wild? Yes. Implausible? After all I experienced in October 2019, I don't dismiss anything.

In 2020, there was a flurry of activity concerning Mars and its myriad mysteries. One of the most eye-opening stories surfaced in January of that year. It all revolved around none other than entrepreneur Elon Musk. As *Business Insider* noted, Musk is planning on sending "one million people to Mars by 2050 by launching three starship rockets every day and creating 'a lot of jobs' on the red planet." The article—written by Morgan McFall-Johnsen and Dave Mosher—states: "Musk said he hoped to build one thousand Starships—the towering and ostensibly fully reusable spaceship that SpaceX is developing in South Texas—over ten years. That's one hundred Starships per year. Eventually, Musk added, the goal is to launch an average of three Starships per day and make the trip to Mars available to anybody."

If anyone can do it, it's surely Musk. Perhaps, it will be Musk and his team—rather than NASA—who discover the truth of life on Mars.

Also in 2020, a particularly intriguing revelation concerning the mysteries of Mars was revealed by Janice Friedman at the *Ancient Code* website. It was a discovery that, in a strange way, blended the distant past and the present day. Friedman wrote, "A new study published by scientists from the Imperial College London suggests there could be as much as 12,000 Olympic sized pools of organic matter on the red planet. After examining Dorset's acidic stream, scientists believe there could be as much as 12,000 Olympic sized pools of organic matter on the red planet.

"Based on analysis of acidic streams in St. Oswald's Bay on the Jurassic Coast in Dorset, England, scientists believe up to 12,000 Olympic sized pools of organic matter may exist on the red planet. Experts say that modern-day streams found in Dorset are similar to Mars's ancient waterways. Scientists studied acidic streams in Dorset because they are 'eerily similar' to the environment on Mars, billions of years ago."

One of the most significant developments in the quest to find life on the red planet hit the headlines on May 8, 2020. The source was *Sci Tech Daily*. It was reported that, "It was already known that there must have been water on Mars, but now the first evidence of rivers in long-term action preserved in exposed cliff faces has been found. 'Rivers that continuously shifted their gullies, creating sandbanks, similar to the Rhine or the rivers that you can find in Northern Italy.' Using high-resolution orbital imagery of the Martian surface, an international team of scientists discovered the stratigraphic product of multiple extensive fluvial-channel belts in an exposed vertical section at Izola Mensa in the northwestern rim of the Hellas Basin. The study recently appeared in the prestigious journal *Nature Communications*."

It was added, "Now Dr. Francesco Salese and Dr. William McMahon along with an international team composed by scientists from

Italy, UK, France, and the Netherlands examined high resolution (25cm/pixel) satellite data of the Hellas (Izola mensa) region to study the characteristics of the newly discovered sedimentary rocks."

An undeniable development in the quest to figure out just how much water Mars once possessed.

"Watch NASA test the Mars Perseverance rover ahead of launch." That was the headline on May 18, 2020 at *Mashable.com*. Journalist Amanda Yeo said, "NASA's Mars 2020 Perseverance rover is getting ready for its big launch day, when it will become the envy of all and leave this cursed Earth behind. But before it can get its ass to Mars, NASA needs to be sure its rover can handle the adventure.

"In a new video filmed in late 2019, NASA's Jet Propulsion Laboratory puts Perseverance through its paces to make sure that it's grown up big and strong. It seems like one giant leap to go from Rover's first unassisted stand to a driving test within just four months, but apparently rovers mature much more quickly than humans. Perseverance's launch window opens July 17. It's scheduled to land on Mars on 18 Feb. 2021, where it will search for aliens and collect cool rocks to show its parents."

Let us hope that the aforementioned "search for aliens" will achieve its goal.

On May 19, *Science Daily* ran an article titled "NASA's Curiosity rover finds clues to chilly ancient Mars buried in rocks." In part, it stated, "Using NASA's Curiosity Rover, scientists have found evidence for long-lived lakes. They've also dug up organic compounds, or life's chemical building blocks. The combination of liquid water and organic compounds compels scientists to keep searching Mars for signs of past—or present—life. Despite the tantalizing evidence found so far, scientists' understanding of Martian history is still unfolding, with several major questions open for debate. For one, was the ancient Martian atmosphere thick enough to keep the planet warm, and thus wet, for the amount of time necessary to sprout and nurture life? And the organic compounds: are they signs of life—or of chemistry that happens when Martian rocks interact with water and sunlight?"

Intriguing questions, to say the very least.

CHAPTER 25

"Martians Stranded [on Earth]"

|||||||||||||||||||||||||||||||||||||

M ac Tonnies was not just known and respected for his work in the field of Martian anomalies: He also had a deep interest in the field of what he personally termed the "Cryptoterrestrials." It's a term he coined in the early 2000s, when his research into this particular arena began. For Tonnies the Cryptoterrestrials fell into two, clearly delineated groups. In essence, it goes like this: For Tonnies, at least some UFO encounters and incidents—particularly alien abductions—were not the work of extraterrestrials. Rather, they were the work of an extremely ancient race of humanoids that developed alongside us, but who chose to stay away from us, aside from when they needed certain things from us—those "certain things" being our DNA, cells, eggs, sperm, and so on—due to the fact that their civilization, today, is degrading and decaying and requires new blood. So, they use us to beef up their race—at least, to the extent that they are able to do so.

Complicating this *already* controversial scenario, Tonnies also speculated on the possibility that there is *another* group of Crypto-terrestrials in our midst. Like that other group he was pursuing,

Tonnies suggested they, too, were equally careful to remain hidden whenever and wherever possible. Tonnies suspected that this second group was possibly Martians—nothing less than the descendants of those earlier Martians who fled their world in an ancient, unclear time, when Mars was facing near-destruction, whether due to war, atmospheric collapse, or both. Tonnies took things to an even more controversial level when he pondered on the scenario of *both groups working together,* in tandem, as a means to save themselves and to protect themselves from us, the admittedly violent and destructive human race. Tonnies wondered what it might be like for the two-tiered Cryptoterrestrials—one of an ancient human type and the other a ragged band of Martians—fighting to live on and having to share the Earth with us, their worst potential enemy possible. Tonnies concluded that it would make good sense for the two factions to band together and carefully mask their real origins and intents.

Tonnies made a very good point when he noted that the two, primary dominating types of aliens that are reported by eyewitnesses are (1) the bug-eyed, insect-like Grays and (2) the very-human-looking Space Brothers of the type that George Van Tassel met out at Giant Rock in the 1950s. Tonnies suggested that the Space Brothers were not aliens, but Cryptoterrestrials. He also opined that the Space Brothers presented themselves in the way they did—as concerned ETs who wanted us to dismantle our weapons of mass destruction—because they knew that if we provoked a third world war, they too, would be annihilated. So, they did what they could to help the situation—particularly in the 1950s, when there was a great deal of alarm and anxiety about nuclear war—by disguising themselves as something very different to their real form.

As for those Martian Cryptoterrestrials, Tonnies felt that they may have been here for so long that they now consider themselves as citizens of the Earth—but, obviously not as humans. At some point, Tonnies speculated, both groups may have agreed to band together, presenting themselves to us as something very different to what they really are: (1) an offshoot of us and (2) a race of stranded Martians whose technology may not be sufficient enough to allow them to return to their home planet of Mars—or what is left of it.

In terms of his theorizing, Tonnies said:

After devouring countless books on the UFO controversy and the paranormal, I began to acknowledge that the extraterrestrial hypothesis suffered some tantalizing flaws. In short, the "aliens" seemed more like surreal caricatures of ourselves than beings possessing the god-like technology one might plausibly expect from interstellar visitors. Like [UFO researcher] Jacques Vallee, I came to the realization that the extraterrestrial hypothesis isn't strange enough to encompass the entirety of occupant cases. But if we're dealing with humanoid beings that evolved here on earth, some of the problems vanish. My hypothesis works too when we apply it to Martians stranded [on Earth] and who, I sometimes wonder, are waiting for the day when our world becomes theirs.[1]

Tonnies continued:

I envision the Cryptoterrestrials engaged in a process of subterfuge, bending our belief systems to their own ends. And I suggest that this has been occurring, in one form or another, for an extraordinarily long time. I think there's a good deal of folkloric and mythological evidence pointing in this direction, and I find it most interesting that so many descriptions of ostensible "aliens" seem to reflect staged events designed to misdirect witnesses and muddle their perceptions.[2]

The one area that really caught the attention of Tonnies, when it came to addressing the possibility that the UFO phenomenon is driven not by aliens but by Cryptotererstrials, was that of the alien abduction phenomenon. There can be very few people—if any—

who have not heard of the controversy, such is the ways and means by which it has filtered into our society's pop-culture and on-screen entertainment. There's no doubt at all that the most famous of all alien abduction incidents was also one of the very earliest: that of Betty and Barney Hill, of New Hampshire, whose lives were turned upside down after a strange and nightmarish encounter on the night of September 1961. Everything was going well for the Hills, for a while anyway.[3]

On the night in question, Betty and Barney were driving home from a vacation in Canada, with a trip to Niagara Falls being a highlight for the pair. If the Cryptoterrestrials are, as Tonnies strongly suspected, impoverished beings—but ones who try their utmost to make it seem that they are our technological superiors—then choosing the location where the incident occurred makes a lot of sense. Probably limited in terms of their abilities to come and go without being seen, the Cryptoterrestrials chose the area carefully: It was a long and winding—and dark—stretch of tree-shrouded road that few people were using that night. A strange light in the sky suddenly appeared, something that made Betty and Barney uneasy. That was hardly surprising, as it seemed like the light—or whatever it really was—was shadowing the Hills, watching them carefully. Eventually, the light went away and the Hills, along with their faithful dog, Delsey, who appears in so many of the Hills's own photographs, continued onward to their home.

Puzzlement turned into deep concern when the Hills realized, on arriving home, that a significant amount of time seemed to have been "lost" during the course of the drive home. As the days and weeks progressed, bad dreams—of having been taken from their car by strange, humanoid figures, and taken aboard a craft and subjected to medical experiments—turned into absolute nightmares. Hypnosis soon followed, and what came out was a story of full-blown alien abduction. In 1966, John G. Fuller wrote a full-length book on the Hill affair, titled *The Interrupted Journey*. One can say that the alien abduction phenomenon was well and truly born. It has never gone away, largely as a result of the coverage given to the subject in Budd Hopkins's 1981 book, *Missing Times,* and Whitley Strieber's bestseller of 1987, *Communion.*

Mac Tonnies had a lot to say about the alien abduction controversy, most of it from a negative perspective when it came to the theory that the perpetrators were real aliens, and wholly positively from the Cryptoterrestrial perspective:

> I regard the alleged "hybridization program" with skepticism. How sure are we that these interlopers are extraterrestrial? It seems more sensible to assume that the so-called aliens are human, at least in some respects. Indeed, descriptions of intercourse with aliens fly in the face of exobiological thought. If the cryptoterrestrial population is genetically impoverished, as I assume it is, then it might rely on a harvest of human genes to augment its dwindling gene-pool. It would be more advantageous to have us believe we're dealing with omnipotent extraterrestrials, rather than a fallible sister species. The ET-UFO mythos may be due, in part, to a long-running and most successful disinformation campaign.[4]

Tonnies had a particular interest in the 1957 abduction of a Brazilian man, Antonio Villas Boas, which he felt was a classic example of a kidnapping at the hands of Cryptoterrestrials. Tonnies said of this still-controversial case, which saw Villas Boas have sex with a very human-looking woman:

> After intercourse, the big-eyed succubus that seduced Antonio Villas-Boas pointed skyward, implying a cosmic origin. But the mere fact that she appeared thoroughly female, and, moreover, attractive, belies an unearthly explanation. Further, one could argue that the clinical environment he encountered aboard the landed "spacecraft" was deliberately engineered to reinforce his conviction that he was dealing with extraterrestrials.

Now on a roll, Tonnies added:

If cryptoterrestrials are using humans to improve their genetic stock, it stands to reason they've seen at least a few of our saucer movies. As consummate anthropologists, they likely know what we expect of "real" extraterrestrials and can satisfy our preconceptions with a magician's skill. Their desire for our continued survival, if only for the sake of our genetic material, may have played a substantial role in helping us to avoid extinction during the Cold War, when the UFO phenomenon evolved in our skies; much to the consternation of officialdom.

Turning his attention to the era of the 1950s—when aliens didn't have big heads and black eyes, but looked far more like 1960s-era hippies—Tonnies had a few profound thoughts that so many UFO researchers failed to address:

Commentators regularly assume that all the Contactees [a term for those who claimed encounters with human-like aliens, chiefly in the 1950s] were lying or else delusional. But if we're experiencing a staged reality, some of the beings encountered by the Contactees might have been real and the common messages of universal brotherhood could have been a sincere attempt to curb our destructive tendencies. The extraterrestrial guise would have served as a prudent disguise, neatly misdirecting our attention and leading us to ask the wrong questions; which we're still asking with no substantial results.

Interestingly, Tonnies wondered if the pale-faced, skinny, mannequin-like creeps known as the Men in Black—who *also* surfaced in the early 1950s—are part and parcel of the Cryptoterrestrials and their agenda. Despite the imagery and storylines that are presented in the phenomenally successful *Men in Black* movies, the fact is that the real MIB are not from the US government. They are not ufological 007s. They don't even look human—and they don't act like us, either. Rather, with their black suits, black fedoras, and gaunt, plastic-like faces, they are clearly not from anywhere around here.

The phenomenon of the Men in Black began in the early 1950s, with a man named Albert Bender, who passed away in 2016 at the age of ninety-four. It was shortly after he created a UFO research group (the International Flying Saucer Bureau [IFSB]) in Bridgeport, Connecticut, in the early 1950s, Bender was visited by a trio of menacing men in dark suits. They were not from the CIA, nor the FBI or the Air Force. The three "men" quite literally materialized in Bender's attic bedroom. They made it abundantly clear that he should leave Ufology behind him. He did: Bender closed down the IFSB and quit Ufology, only returning briefly in 1962 to write a book on his experiences. Its appropriate title was *Flying Saucers and the Three Men*. Bender spent the rest of his life living quietly with his wife and family in Los Angeles, California. Since the days of Albert Bender there have been hundreds of reports of disturbing encounters with the MIB. Threats from the Cryptoterrestrials? Tonnies was sure he was on the right track.

That Tonnies's theory was so controversial, alternative, and almost unique—Martians and ancient humans working together to save themselves, and to hell with us when they are done with the human race—inevitably ensured that Tonnies would get some flak. And, guess what? He did. Nevertheless, Tonnies did have more than a few points in his favor. On this particular point, he said to me:

> The cryptoterrestrial theory has met with mixed reactions. Some seem to think that I'm onto something. Most UFO researchers are, at best, extremely skeptical. Others think I'm parroting John Keel's "superspectrum" [Keel was the author of the acclaimed 1975 book, *The Mothman Prophecies*], a variation on the "parallel worlds" theme

that in turn shares memes with Jacques Vallee's "multiverse." Both ideas suggest that we somehow occupy dimensional space with our "alien" visitors, doing away with the need for extraterrestrial spacecraft while helping explain the sense of absurdity that accompanies many UFO and occupant sightings. Keel and Vallee have both ventured essentially "occult" ideas in cosmological terms; both the "superspectrum" and the "multiverse" require a revision of our understanding of the way reality itself works. But the crypto-terrestrial hypothesis is grounded in a more familiar context.

He added:

I'm not suggesting unseen dimensions or the need for ufonauts to "downshift" to our level of consciousness. Rather, I'm asking if it's feasible that the alleged aliens that occupy historical and contemporary mythology are flesh-and-blood human-like creatures that live right here on Earth. Not another version of Earth in some parallel Cosmos, but on Earth. While I can't automatically exclude the UFO phenomenon's "paranormal" aspects, I can attempt to explain them in technological terms. For example, I see no damning theoretical reason why "telepathy" and "dematerialization" can't ultimately be explained by appealing to cybernetics, nano-technology, and other fields generally excluded from ufological discourse.

In finality, Tonnies provided these words: "The cryptoterrestrial hypothesis manages to alienate champions of the extraterrestrial hypothesis and those who support a more esoteric, 'inter-dimensional' explanation. It offers no clear-cut reconciliation. It does, however, wield explanatory potential lacking in both camps."

It's not just Tonnies who has pondered on this particularly intriguing scenario, as you'll see now.

A long-time figure in Ufology, Timothy Green Beckley has taken a look at the Cryptoterresrial scenario, too:

Some people might suggest that because the Space Brothers look so much like us that they could be from somewhere right here on Earth—an ancient race, maybe. There are a lot of cultural legends about advanced beings living underground: UFOs coming out of the oceans, lakes, caverns. The whole hollow-Earth thing is a little hard for me; but the caverns theory I can take.

I remember one incident, 1970s, where I was lecturing and a gentleman—a professor at the college where I was lecturing—came up to me and told this story about how he was driving outside of a town in Michigan. It was rather late at night, and he saw these lights in the woods. He pulled over, and there was no other traffic coming in either direction; but there was already another car parked at the side of the road.

He described seeing some sort of ship in the distance—a UFO. A group of human-like aliens got out, walked to the car, which was a Cadillac, or something like that. He watched them and could see they looked human. They just got in the car and drove off. But then, a couple of weeks later, he sees one of the same guys in a supermarket. These reports sound far-fetched; but there's so many of them of what seem to be aliens being able to move among us. But, if they're really from here, that might explain it.[5]

What about the Cryptoterrestrials, Martians, and elements of the US government? Is there a conspiracy within the government to hide the truth of these ancient beings? There just might be. Walter Bosley is a successful writer of mysteries and esoteric topics who goes under the alias of E. A. Guest. Bosley has an interesting background, having been in the employ of both the FBI and the US Air Force's Office of Special Investigations. It's not to Bosley, directly, we have to turn our attentions. Rather, it's his father.

In the 1950s, Bosley's father worked on certain classified programs that benefited the pre-NASA US space program, much of it in relation to flight medicine training. On one memorable occasion, Bosley Sr. was ordered to make a flight to Wright–Patterson

Air Force Base, located in Dayton, Ohio. It was a day that Walter Bosley's father would never forget: He was briefed on what really happened near Roswell, New Mexico, in early July 1947—the most famous UFO case of all. Bosley's father was told that nothing of an extraterrestrial nature came down in the wilds of New Mexico. Walter Bosley said:

> According to my father, these vehicles came from inside the planet. The civilization . . . exists in a vast, underground system of caverns and tunnels beneath the southwest and is human. Occasionally, they come and go, emerging in their vehicles and occasionally they crash. They are human in appearance, so much so that they can move among us with ease [and] with just a little effort. If you get a close look, you'd notice something odd, but not if the person just passed you on the street.[6]

Notably, Bosley added: "I believe that the ET hypothesis has been used by the 'aliens' themselves, because it is most readily embraced by people who have had encounters with them."[7]

Mac Tonnies contemplated on this particularly fascinating and alternative theory for what happened in '47:

> The device that crashed near Roswell in the summer of 1947, whatever it was, featured properties at least superficially like the high-altitude balloon trains ultimately cited as an explanation by the Air Force. Debunkers have, of course, seized on the lack of revealingly "high-tech" components found among the debris to dismiss the possibility that the crash was anything but a case of misidentification; not even Major Jesse Marcel, the intelligence officer who advocated an ET origin for the unusual foil and structural beams, mentioned anything remotely resembling an engine or power-plant. The cryptoterrestrial hypothesis offers a speculative alternative: maybe the Roswell device *wasn't* high-tech. It could indeed have been a balloon-borne surveillance device brought

down in a storm, *but it doesn't logically follow that is was one of our own* [author's emphasis]. Upon happening across such a troubling find, the Air Force's excessive secrecy begins to make sense.[8]

Are impoverished Martians secretly working with an ancient human race, as a means to try and steer us away from nuclear war? At the time of his death in 2009, Tonnies was pretty sure that he was at least heading in the right direction.

Conclusion

|||||||||||||||||||||||||||||||||||||||||

Our investigation into the strange saga of Mars, its many anomalies, and its mystery-shrouded past, is over. At least, it's over to the extent that we, from right here on Earth, can investigate the controversy. That's to say from the perspective of a world that is millions of miles away from us. While there is no doubt that there are mountains of questions and unresolved issues concerning the Red Planet, the likelihood is that we will never know all of the answers to those questions until we make a manned journey to Mars and construct permanent facilities to allow for studies of Mars on a 24/7 basis. Certainly, more than a few probes sent to Mars have done exactly that over the years and decades. Having a wholly human-driven perspective on matters of Martian mysteries, however, is vital. How long that might take is, right now, something that we cannot ascertain. After all, May 2020 was the first time NASA astronauts were sent into Earth orbit in almost a decade, thanks to Elon Musk's SpaceX. For many, that might be seen as a big problem. On the other hand, it doesn't really matter which nation sends a crew to Mars first: just that *someone* does. And when that happens, we will hopefully know if there was once intelligent life on Mars—highly intelligent life, too.

Perhaps with bases on Mars we will be able to confirm that yes, there once was a culture on the Red Planet that flourished, but that finally came to an end. On the other hand, maybe some of the answers will lead us down the roads marked "Pareidolia." Such a scenario is not at all impossible. And, if that is where the answers are, well, that will say far more about us as a species, and our yearnings (subconscious or otherwise) to find evidence that we are not all alone in the universe. With that all said, let's now see where things are presently.

It's important to note that Mars has an abundance of water, most of it in the form of ice, but significant amounts in liquid form, too. On the latter point, we should take close and careful note of the words of Philip Ball, a consultant editor for *Nature*. He says on the matter of life and water:

> At a somewhat crude level, life is about molecular processes, such as templating, molecular recognition, and replication, which can be duplicated in non-aqueous solvents. But we find that even for the simplest organisms, many of the molecular interactions are facilitated by water in an extremely fine-tuned way. I'm not sure we know of any solvent that can play a comparable role in terms of enabling the kind of highly delicate chemistry that makes life possible. If we accept that any form of life will require a comparable degree of chemical sophistication, it is hard to see what other solvent would make this possible.[1]

We would do well to consider Ball's words very carefully, as we know that Mars is teeming with water. So, when it comes to the matter of life on Mars, if nothing else it's a world that has a significant amount of one of the things that we cannot live without: water. That's to say Mars is in a far better position to harbor life than, let's say, Mercury. Charles Q. Choi says of Mercury:

> Because the planet is so close to the sun, Mercury's surface temperature can reach a scorching 840 degrees Fahrenheit (450 degrees Celsius). However, since this world doesn't have much of a real atmosphere to entrap any heat, at night temperatures can

plummet to minus 275 F (minus 170 C), a temperature swing of more than 1,100 degrees F (600 degree C), the greatest in the solar system.[2]

It must be said that Mars hardly has a pleasant environment, but it is at least a step ahead of not just Mercury, but Venus too—a planet that is so hot, lead will melt on its surface. Never mind running water, Venus is almost completely lacking in water vapor. Current estimates show that the atmosphere of Venus consists of a minute 0.002 percent water vapor. Does all of this make Mars the most likely place where we might find life? Yes, it does. We should also keep in mind the matter of those strange and amazing "banyan trees" that sci-fi writer Arthur C. Clarke championed as evidence of plant life on Mars. Clarke provided a solid reason as to why he was right in his belief. NASA, however, had its very much down-to-earth explanation, as we've seen. Clarke is gone, but the debate on the banyan trees is certainly not.

Now, keeping all of the above in mind, let's address the primary controversies that this book offered. We'll begin with the matter of comic-book legend Jack Kirby and the acclaimed author of *Gulliver's Travels*, Jonathan Swift. The stories of both men, specifically as they relate to Mars, suggest that the two had secret knowledge concerning both the planet and its distant past. In terms of Swift, the controversy revolved around Mars's moons, Phobos and Deimos. As for Kirby, the topic was focused on his 1950s-era comic book story, "The Face on Mars," as well as a similar story, "The Great Stone Face." Did both men know of ancient, hidden information concerning Mars and its civilization? Maybe they did. After all, and as we've seen, Kirby moved in intriguing circles. And, as we have also seen, in a roundabout fashion Kirby had ties to the CIA. And, as we can say for sure, thanks to the provisions of the US government's Freedom of Information Act, in 1984 the CIA secretly remote-viewed Mars and secured extraordinary data of the planet, of its "people," and of its long-gone civilization. Threads that are a vital part of the Martian mysteries? Or, on the other hand, was all of this just a case of coincidence piling upon coincidence? That's the frustrating trend that permeates throughout the pages of this book: that we have some truly tantalizing data, but nothing that can be seen as 100 percent solid. As the author, I have to admit that. When it comes to Kirby and Swift, the true believers will continue to believe, while the debunkers will continue to debunk. And, in the process, neither camp, so far, has achieved a damned thing.

What of the strange "objects," "artifacts," and other extraordinary "things" on the surface of Mars that have been photographed and filmed by a variety of NASA's rovers and probes? This is where things become even more and more controversial. There can be absolutely no doubt whatsoever that it was the discovery of the "Face on Mars" in the 1970s that energized the interest and controversy that surrounded (and that still continues to surround) the possibility that Mars may have been a world not too dissimilar to ours, even a world inhabited by beings who resembled us to a staggering degree—aside, that is, from their alleged immense height.

Add to that the D&M Pyramid and the "Sphinx" that Mike Bara champions, and what we have is not just a possibility that the architecture of the Martians resembled much of ours, but a theoretical connection to Egypt millennia ago—a sensational scenario worthy of the History Channel's *Ancient Aliens* show. Again, it's a collective body of data that tantalizes us, and it provokes images of Martians visiting (or fleeing to) Earth to start a new life for themselves, this latter point speculated on by Mac Tonnies. He offered that a small, impoverished body of Martians not only arrived on our world in the past and, possibly, were responsible for our beliefs in a god or in multiple gods among us, but who may still be with us today—in deep stealth, of course. Again, the scenario is fascinating, but it's inconclusive, mainly because we are presently unable to venture to Mars and tread the landscape where these anomalies can be found.

Then, there are the *really* controversy-filled issues, such as that image of a cat-like (or an Anubis-like) animal on Mars that Mike Bara has highlighted in his work; the "Marshenge"; Tesla's suspicions that he had contact with Martians; the monoliths on both Mars and Phobos; the weird saga of Mr. Axelrod and his apparent pressing, in 1975, to have the Moon remote-viewed, something that the CIA did a handful of years later, but in relation to Mars; the alleged connections between ancient nuclear war (on both Mars and Earth); and the presence of the Anunnaki in all of this.

Each and every one of them is just about guaranteed to amaze us and provoke debate—and, let us be clear, to entertain people, too. Once again, we're in that domain where fact, fiction, speculation, conspiracy-theorizing, and the important work of NASA are all rolled into one—and sometimes in chaotic fashion. That does not, however, exclude the possibility that what we are seeing on the photos coming back from Mars are exactly what many investigators of the Martian mysteries believe them to be: evidence of a Martian culture way back in the fog of time. There is also the matter of conspiracies relative to Mars.

The fact that in 1984 the CIA chose to secretly remote-view Mars in the period of 1,000,000 BC, strongly suggests that someone—or some group/think-tank—within the agency had a deep interest in not just Mars's distant past, but its civilization, its citizens, and its ultimate, violent collapse. Why? Well, that question alone most certainly *does* provoke thoughts and images of conspiracy and cover-up. Of what? We don't know. On the other hand, I find it difficult to believe that there is a huge, all-encompassing, secret program hidden deep specifically within NASA, the role of which is to hide the truth of a mighty, Martian empire that once ruled the Red Planet. Why do I doubt? It's quite simple.

The only reason why we know that there appear to be strange things on Mars is because NASA has made the specific, relevant photographic evidence freely available to us! Whose technology photographed the Face on Mars, the D&M Pyramid, the Stone-henge-like "Marshenge," the monoliths on Mars and Phobos, and the Sphinx? I'll tell you: NASA's technology, not that of crusading UFO researchers. And, it was NASA who placed those same images into the public domain and with easy access and at multiple web-sites. One cannot have it both ways: Either NASA is hiding some-thing or it isn't. Unless, that is, you take the view that the release of all of these pictures is part of a bigger program to slowly, and carefully, reveal "the truth" to us, one and all. Admittedly, that could be *exactly* what's going on; it will, however, require harder evidence than we have right now to say that disclosure is looming large on the horizon.

Although I don't see NASA engaged in a huge cover-up of the Martian structures, what I *do* see is a mistake on its part. That mis-take has been to ignore and summarily dismiss the many unde-niable oddities away as the results of nothing but pareidolia. The purpose behind all of this lack of interest or involvement by the space agency? Probably NASA's fear of its reputation being besmirched, and the risk of its staff relegated by the media and much of the scientific community to "UFO/alien hunters." A great deal more, by now, could have been achieved by NASA to either bring the whole controversy to rest in a down-to-earth fashion, or to admit that certain phenomena on Mars deserve to be scrutinized to a greater degree—even if doing so might provoke risible comments and criticisms directed at NASA.

If it looks like and sounds like I'm fence-sitting here, I'm actu-ally not. My view on all of this goes as follows: I believe that at least *some* of the Martian curiosities filmed and photographed by NASA probably *are* evidence that we are not alone in the universe. In fact, we may have had alien cousins who were, in terms of distance, astonishingly close to us way back when—and who finally landed and possibly even intervened. Possibly they still are, if one buys into the theories concerning Mac Tonnies's Cryptoterrestrials. Now, moving onto the end, in more ways than one . . .

Not at all unlike what we see in the final moments of the 1968 movie *Planet of the Apes,* the Martian landscape appears, to me, to be littered with evidence of what was once at least one or several sprawling cities, but all now rendered pulverized and flattened. And a civilization collapsed. In the movie, such a city would turn out to be New York, as the presence of a ruined and pummeled Statue of Liberty makes abundantly and graphically clear. That's when Charlton Heston's astronaut character, Taylor, learns to his sudden horror that he is not on a faraway world, after all, but on Earth, thousands of years after a nuclear war has all but decimated our world and its people.

There's another parallel with that classic movie, too: One of the primary characters in *Planet of the Apes* is Dr. Zaius, the minister of science. He has spent much of his life terrified by the possible revealing of the true history of the ape planet—namely, that we, the human race, came first and that, in terms of technology, we were once far ahead of the apes' society. Zaius's fear is that opening the equivalent of Pandora's box will cause chaos in the world of the apes. So, he takes the only option he thinks is viable: Zaius buries everything, just in case, even if he's not fully sure of what went on before his civilization emerged.

I sometimes ponder on the possibility that some of NASA's staff, not unlike Dr. Zaius, might be fearful of what the ramifications could be if history is changed by the opening of that aforementioned box. So, rather like the person who wakes up one morning and finds a lump under one of their armpits, and who refuses to go to the doctor for fear of what they might learn, the truth is ignored, with a hope that it will go away. But it doesn't go away. It just piles up more and more. Maybe, that's how things are with NASA: a case of "If we forget about the Face on Mars we won't have to deal with it." But, it's clearly not going away. Nor are the many other anomalies that are scattered across the Red Planet.

Finally . . .

As a species, we are woefully unaware, and largely ignorant, of a series of incredibly ancient events that revolved around a nightmarish, irreversibly collapsing Mars. And of death on an almost incomprehensible scale for the Martians, most of them killed before they could flee their world for ours. When? Without doubt millennia ago—probably even way back further than that. As I noted earlier in the pages of this book, trying to place together a definitive, smooth timeline for when the Martians lived and died, when the Anunnaki chose to interfere in matters, when a war on Mars may have spilled over to the Earth—hence the stories of Sodom and Gomorrah—and when so much damage was done to the D&M Pyramid and the Face on Mars is without doubt impossible. At best, the timeline is absolutely chaotic in the extreme. As for that ignorance and unawareness I refer to above, in my view it's solely due to the passage of time, to the distortion of incredible history to fanciful mythology, and to legend mutated into something that largely most people write off as being just too good to be true.

Eerie parallels with 1968's Planet of the Apes. (Michael Pye)

When it comes to the scientific community, it is almost exclusively unwilling to address the mysteries of Mars in a fashion that just might have a potentially adverse bearing on their lofty, much-cherished reputations. Some, I suspect, are fearful of what they might find if they dare to go looking too closely—what we might call a skewed case of deliberate self-censorship. That point is something that *also* harks back to *Planet of the Apes*. As the movie comes to its chilling, unforgettable ending, Dr. Zaius cautions Charlton Heston's character, Taylor, with these words: "Don't look for it, Taylor. You may not like what you find."

Notably, there is evidence to support this particular angle of stumbling on something disturbing in the extreme and that will be life-changing for all of us and related to life on Mars. On September 30, 2019, CNN ran a story with the following title: "When—or if—NASA Finds Life on Mars, the World May Not Be Ready for the Discovery, the Agency Chief Says." And, as it turned out, this was not an overly sensationalized story. It was prompted by the words of NASA's chief scientist, Jim Green. He said: "It will be revolutionary. It will start a whole new line of thinking. I don't think we're prepared for the results. We're not."[3]

Green continued: "What happens next is a whole new set of scientific questions. Is that life like us? How are we related?"[4] There's no doubt that Green's words are astonishing—taking into consideration who he is and his stature within NASA. Green's statement also had somewhat of an ominous tone attached to it; of that, it's very hard to deny.

In time, perhaps we'll all find out why, exactly, "the world may not be ready" for what NASA *might* find or already *has* found on Mars. Let's hope the news from the Red Planet is good.

Acknowledgments

I would like to thank everyone at Red Wheel Weiser and particularly Greg Brandenburgh, Michael Pye, Laurie Kelly, Bonni Hamilton, Michael Kerber, Eryn Carter, Michelle Spanedda, Kathryn Sky-Peck and Jane Hagaman; my literary agent and good friend, Lisa Hagan, for her tireless work; Linda Godfrey for a great interview; and Kimberly Rackley, without whose expertise in the field of remote viewing significant portions of this book could never have been written.

Chapter Notes

Introduction
1 "Early Times."
2 Gill, "Mars."
3 Wigington, "Mars."

Chapter 1: "The Human Race Has Something to Do with Mars"
1 Tonnies quotes throughout this section from Redfern, interview with Mac Tonnies, March 14, 2004.
2 Redfern, interview with Mac Tonnies, September 9, 2006.

Chapter 2: "Two Lesser Stars or Satellites, which Revolve about Mars"
1 Swift, *Gulliver's.*
2 Ibid.
3 Darling, "Jonathan."
4 Ibid.

Chapter 3: "This Is a Message from Another Planet, Probably Mars"
1 "Airships."
2 Colavito, "A Message."
3 "Message Perhaps."
4 Ibid.
5 Ibid.
6 "Wiggins."
7 "Is This Meteor?"
8 "Wiggins."
9 Ibid.
10 All quotes throughout this section are from "Is This Meteor?"
11 "Notes and News."
12 Bell, "Things."

Chapter 4: *"Mars As the Abode of Life"*
1 "NASA – Report."
2 Wagner, "The Most."
3 Nicks, "Opening."

4 Hoagland and Bara, *Dark.*
5 Hiemstra, "The Glass."
6 Wagner, "The Most."
7 Redfern, interview with Mac Tonnies, September 9, 2006.

Chapter 5: "Is Mars Sending Us Signals?"
1 "World Speaks."
2 Ibid.
3 Quotations throughout this section, unless otherwise noted, come from "Nikola Tesla," which includes multiple quotations from interviews of the day.
4 Rux, *Architects.*
5 Hester, "Everybody."
6 "Marconi."
7 Ibid.

Chapter 6: "Objects Sighted May Possibly Be Ships . . . from Mars"
1 This and the following quotations from Arnold are from "Project 1947."
2 "Desmond."
3 "UFO."
4 "George Van Tassel."
5 "The Immortality."
6 "George Van Tassel."
7 Ibid.

Chapter 7: "A Visual History of a Race's Heroic Death"
1 Knowles, "Mindbomb."
2 Ibid.
3 "Biography."
4 "Did You Know?"
5 "Space-Life Report."
6 Ibid.
7 Ibid.

Chapter 8: "They're Ancient People. . . . They're Dying"
1 Smith, "What?"
2 Kress, "Studies."
3 Ibid.
4 "Mars Exploration."
5 Ibid.
6 Albarelli, *A Terrible.*

Chapter 9: "Strange Geometric Ground Markings and Symbols"
1 Swann, *Penetration*; "Ingo Swann"; Sprague, "Third Eye"; and Wagner, "What's?"
2 Ibid.
3 Ibid.

4 Leonard, *Somebody.*
5 Hoagland, "Preliminary."

Chapter 10: "Unusual Images Were Radioed Back to Earth"

1 "Mars Observer."
2 Evans, "And Then."
3 Hoagland and Bara, *Dark Mission.*
4 Oberg, "The Dark."
5 Ibid.
6 Strauss, "Ten."
7 Oberg, "The Dark."
8 Lee, "Did NASA"?
9 "The Complete."
10 "Reexamining."
11 "Phobos: Malfunction."

Chapter 11: "The Image Is So Striking"

1 Foulke, "The Banyan."
2 "Mars Global."
3 Ibid.
4 Ibid.
5 Ibid.
6 Nelson, "9 Features."
7 Space.com, "Are Those?"
8 Foulke, "The Banyan."
9 "Dalmatian."
10 Redfern, interview with Mac Tonnies, March 14, 2004.
11 Greicius, "Jamming."
12 Steigerwald, "Designer."
13 Jordan, "Can Plants?"

Chapter 12: "We Should Visit the Moons of Mars. There's a Monolith There."

1 "Martian Moons."
2 Redfern, interview with Mac Tonnies, March 14, 2004.
3 Hille, "Mars' Moon."
4 Wall, "How."
5 Redfern, interview with Mac Tonnies, March 14, 2004.
6 "Martian Moons."
7 "Phobos Facts."
8 Palermo, "Palermo's."
9 Ibid.
10 Hanks, "Spaceships."
11 "Buzz Aldrin."

Chapter 13: "A Beacon Erected by Aliens for Mysterious Reasons"

1 Wolchover, "'Monolith.'"
2 Ibid.
3 "Has the Mystery?"
4 Ibid.
5 "HiRISE."
6 "Has the Mystery?"

Chapter 14: "Has Stonehenge Been Found on Mars?"

1 "Talk: Selim Hassan."
2 "Robert M. Schoch: About."
3 "Robert M. Schoch: Research."
4 Ibid.
5 Mark, "Nefertiti."
6 "Mars Artifacts."
7 Tonnies, *After.*
8 "Twin Peaks."
9 Redfern, interview with Mac Tonnies, March 14, 2004.
10 Bara, *Ancient Aliens on Mars.*
11 Hoagland and Bara, *Dark Mission.*
12 Bara, *Ancient Aliens on Mars.*
13 Redfern, interview with Mac Tonnies, March 14, 2004.
14 Ibid.
15 Austin, "Has Stonehenge?"
16 Ibid.
17 Gray, "Stonehenge-Style."

Chapter 15: "A German Shepherd–Like Head"

1 Sessions, "What Is?"
2 Armstrong, "Pareidolia."
3 "Mike Bara Gets."
4 "Anubis."
5 Klimczak, "Anubis." The source for the colleague who shared his views on Bara's claims is an October 1, 2019, interview.
6 Redfern, interview with Linda Godfrey.

Chapter 16: "It May Be a Crab-Like Animal"

1 Waugh, "Crab-Like."
2 Ibid.
3 Maynard, "Crab-Like."
4 Bara, *Ancient Aliens on Mars.*
5 Ibid.
6 Radford, "Female."
7 Bara, *Ancient Aliens on Mars II.*

Chapter 17: "Liquid Water Flows Intermittently on Present-Day Mars"

1 Anderson, "NASA."
2 "Found!"
3 "Odyssey."
4 "Volcanic Rock."
5 "Mars Atmosphere."
6 Quotations throughout this section are from Halton, "Liquid" unless otherwise noted.
7 Northon, "NASA."
8 Ibid.

Chapter 18: "The Legend of Levitation"

1 Jones and Flaxman, "Our Sonic."
2 Jessup, *The Expanding.*
3 Ibid.
4 Mooney, *Colony.*
5 All quotations in this paragraph are from Collins, *Beneath.*
6 "Stonehenge Bluestones."
7 Jones and Flaxman, "Our Sonic."

Chapter 19: "Ancient Martian Civilization Was Wiped Out"

1 McDaniel and Paxson, The Case.
2 "Dr. John."
3 O'Callaghan, "Ancient."
4 Ibid.
5 Tonnies quotes throughout this section, unless otherwise noted, are from Tonnies, *After.*
6 Redfern, interview with Mac Tonnies, March 14, 2004.
7 Rux, *Architects.*
8 Ibid.
9 Redfern, interview with Mac Tonnies, March 14, 2004.
10 "What Was?"
11 To avoid using too many quotations (because of the old-style language), I put the story together from the following sources: "Assyrian"; Haines, "Assyrian"; Lean, "Sodom"; Longnecker, "Why?"; "Sodom"; and "What Was"?
12 Haines, "Assyrian."
13 Ibid.
14 Ibid.
15 Lean, "Sodom."
16 Ibid.
17 "Human Rights."
18 Ibid.

19 Sitchin, *Genesis.*
20 Hoffman, "Ace."

Chapter 20: "Massive Species Extinctions"
1 "The Atomic."
2 Bennett, "Here's."
3 Ibid.
4 Hertsgaard, "Mikhail."
5 "Nuclear Winter" (abyss.uoregon.edu).
6 Lamb, "What?"
7 Committee, *The Effects.*
8 Ibid.
9 Robock, "Nuclear."
10 Ibid.
11 Baum, "The Risk."
12 "Nuclear Winter" (atomicarchive.com).

Chapter 21: "High-Velocity Impacts"
1 Lonar Crater, India."
2 Ibid.
3 Ramachandran, "Lonar."
4 Childress, *Technology.*
5 Ibid.
6 Benedict, "The First."
7 Firestone and Topping, "Terrestrial."
8 "New Evidence."
9 Ibid.
10 Ibid.
11 Ibid.
12 Taylor, "The Australasian."
13 "The Story."
14 "The Stealth."

Chapter 22: "The Reconstruction of the Devastated Earth"
1 Sangster, "Who?"
2 Ibid.
3 "The Nephilim."
4 "Numbers 13:31."
5 Mark, "Gilgamesh."
6 Sitchin, "In the News."
7 Ibid.
8 *The Epic.*
9 Ibid.

10 "Gilgamesh Tomb."
11 Ibid.
12 Mills, "20 New."
13 "Mars Exploration."

Chapter 23: "They Created a Permanent Space Base on Mars"
1 Redfern, *Bloodline.*
2 Gardner, *Genesis.*
3 "White Powder."
4 "Repairing."
5 Marrs, *Our Occulted.*
6 Ibid.
7 Sitchin, *When Time.*
8 Hardy, *Wars.*
9 Hardy, *DNA.*

Chapter 24: "There Are Thunderous Vibrations that Shake Mars"
1 Rackley quotes throughout are from Redfern, interviews with Kimberly Rackley, October 4, 7, 9, and 10, 2019.
2 "Asteroids."
3 "The 11:11."
4 "11:11."

Chapter 25: "Martians Stranded [on Earth]"
1 Redfern, interview with Mac Tonnies, July 7, 2009.
2 Ibid.
3 The following account has been widely reported but was sourced here by Fuller, *The Interrupted.*
4 Tonnies quotes throughout here, unless otherwise noted, are from Redfern, interview with Mac Tonnies, July 7, 2009.
5 Redfern, *Contactees.*
6 Guest, "The Other."
7 Ibid.
8 Redfern, interview with Mac Tonnies, July 7, 2009.

Conclusion
1 Watanabe, "Water."
2 Choi, "Planet."
3 Andrew, "When."
4 Ibid.

Bibliography

"Airships in America." How Stuff Works website 2019. *science.howstuffworks.com.*

Albarelli, Jr., H. P. *A Terrible Mistake: The Murder of Frank Olson and the CIA's Secret Cold War Experiments.* Walterville, Oregon: Trine Day LLC, 2009.

Alouf, Michel M. *History of Baalbek.* San Diego, California: Book Tree, 1999.

"Ambrosia." Greekmythology.com, 2016. *www.greekmythology.com.*

"Ambrosia—Food of the Greek Gods." Loggia, 2016. *www.loggia.com.*

"Amrita—Ambrosia—The Nectar of Immortality." Yoga with Evie website, May 14, 2013. *yogakinisis.wordpress.com.*

"Ancient Atomic Knowledge?" Biblioteca Pleyades website, 2015. *www.biblioteca pleyades.net.*

"Ancient Chinese Alchemists and Their Search for Immortality." 2016. *www .monkeytree.org.*

Anderson, Gina. "NASA Confirms Evidence that Liquid Water Flows on Today's Mars." NASA.gov, September 28, 2015. *www.nasa.gov.*

Andrew, Scottie. "When—or if—NASA Finds Life on Mars, the World May Not Be Ready for the Discovery, the Agency Chief Says." CNN.com, September 30, 2019. *www.cnn.com.*

"Anubis." Crystalinks, 2019. *www.crystalinks.com.*

"Anunnaki." Halexandria website, updated February 5, 2009. *www.halexandria.org.*

Armstrong, Katherine. "Pareidolia: The Science behind Seeing Faces in Everyday Objects." Lenstore, January 23, 2019. *www.lenstore.co.uk.*

"Assyrian Clay Tablet Points to 'Sodom and Gomorrah' Asteroid." Isegoria website, March 31, 2008. *www.isegoria.net.*

"Asteroids." NASA Science. *solarsystem/nasa.gov.*

"The Atomic Bombings of Hiroshima and Nagasaki." The Avalon Project website, 2008. *avalon.law.yale.edu.*

Austin, Jon. "Has Stonehenge Been Found on Mars? Ancient 'Alien' Stone Circle Discovered on Red Planet." *Express,* September 24, 2015. *www.express.co.uk.*

"Baalbek." Ancient Aliens Debunked, 2015. *ancientaliensdebunked.com.*

Bara, Mike. *Ancient Aliens on Mars.* Kempton, Illinois: Adventures Unlimited Press, 2013.

———. *Ancient Aliens on Mars II.* Kempton, Illinois: Adventures Unlimited Press, 2014.

———. "Mars Pathfinder Landing Site . . . A Sphinx Revisited?" Enterprise Mission, 2001. *www.enterprisemission.com.*

Baum, Seth. "The Risk of Nuclear Winter." Federation of American Scientists, May 29, 2015. *fas.org.*

Bell, Walter George. "Things That Fall from the Sky." *Windsor Magazine* volume 22, 1905.

Benedict, Tim. "The First Global Nuclear War and a Coverup of HISTORICAL Proportions!" AncientNuclearWar.com, 2015. *ancientnuclearwar.com.*

Bennett, Jay. "Here's How Much Deadlier Today's Nukes Are Compared to WWII A-Bombs." *Popular Mechanics,* October 10, 2016. *popularmechanics.com.*

"Biography of Wernher von Braun." NASA website, August 3, 2017. *www.nasa.gov.*

Brandenburg, John E. *Death on Mars: The Discovery of a Planetary Nuclear Massacre.* Kempton, Illinois: Adventures Unlimited, 2015.

Brenna, Herbie. *Martian Genesis: The Extraterrestrial Origins of the Human Race.* New York: Dell Publishing, 1998.

"Buzz Aldrin: We Should Visit the Moons of Mars—There's a Monolith There." Disclose TV, 2019. *www.disclose.tv.*

Caidin, Martin. *Destination Mars: The Next Step in Man's Exploration of Space: The Red Planet by the Noted Aviation and Space Writer.* New York: Doubleday & Company, Inc., 1972.

Canright, Shelley (Ed.). "The 'Canali' and the First Martians." NASA, April 13, 2009. *www.nasa.gov.*

Carlotto, Mark J. *The Martian Enigmas: The Face, Pyramids and Other Unusual Objects on Mars.* Berkeley, California: North Atlantic Books, 1991.

"The Case of the Evil Wind: Climate Study Corroborates Sumer's Nuclear Fate." The Official Website of Zecharia Sitchin, November 2001. *www.sitchin.com.*

Central Intelligence Agency (CIA) files on remote viewing, declassified under the terms of the Freedom of Information Act.

Childress, David Hatcher. *The Anti-Gravity Book.* Kempton, Illinois: Adventures Unlimited Press, 2003.

———. *Technology of the Gods.* Kempton, Ill.: Adventures Unlimited Press, 2000.

———. *Vimana Aircraft of Ancient India & Atlantis.* Kempton, Illinois: Adventures Unlimited, 1988.

Choi, Charles Q. "Planet Mercury: Facts About the Planet Closest to the Sun." Space.com, October 14, 2017. *www.space.com.*

Clark, Gerald R. *The Anunnaki of Nibiru.* CreateSpace independent publishing platform, 2013.

Colavito, Jason. "Ancient Atom Bombs: Fact, Fraud, and the Myth of Prehistoric Nuclear Warfare." 2015. *www.jasoncolavito.com.*

———. "A Message from Mars." 2019. *www.jasoncolavito.com.*

Collins, Andrew. *Beneath the Pyramids: Egypt's Greatest Secret Uncovered.* A.R.E. Press, 2009.

Committee on the Atmospheric Effects of Nuclear Explosions. The Effects on the Atmosphere of a Major Nuclear Exchange. Washington, DC: National Academies Press, 1985.

"The Complete Phobos 2 VSK Data Set." Planetary.org, 2019. *www.planetary.org.*

"Crater Lake: Like No Place Else on Earth." National Park Service website, 2015. *www.nps.gov.*

"Crown Face." Mars Anomalies, 2019. *www.marsanomalies.com.*

"Dalmatian Terrain." Jet Propulsion Laboratory website, June 12, 2003. *www.jpl.nasa.gov.*

Darling, David. "Jonathan Swift and the Moons of Mars." David Darling Info, 2019. *www.daviddarling.info.*

"Desmond Leslie – Sifra di-Tzeniuta, the 'Book of Dzyan,' and 'Flying Saucers Have Landed'." DavidHalperin.net, December 29, 2016. *davidhalperin.net.*

"Did You Know that Walt Disney Helped the U.S. Military Imagine the Experience and Reality of Space Travel?" This Day in Disney History website. *thisdayindisneyhistory.homesteadcom.*

"Dr. John Brandenburg." The Space Show website. *thespaceshow.com.*

"Early Times." Mars Exploration—NASA, 2019. *mars.nasa.gov.*

"The Einstein Factor." HISTORY Canada website, 2015. *www.history.ca.*

"11:11: Is it Happening to You?" Power of Positivity website, 2019. *www.powerofpositivity.com.*

"The 11:11 Phenomenon." Dimension 1111 website, 2011. *www.dimension1111.com.*

"Enki and Ninhursag." Gateways to Babylon, 2015. *www.gatewaystobabylon.com.*

"The Epic of Gilgamesh." SparkNotes, 2015. *www.sparknotes.com.*

The Epic of Gilgamesh. London: Penguin Classics, 2003.

"Epic of Gilgamesh—Sumerian Flood Story 2750–2500 BCE." HistoryWiz, 2015. *www.historywiz.com.*

Estrada, Andrea. "Study Jointly Led by UCSB Researcher Finds New Evidence Supporting Theory of Extraterrestrial Impact." The Current, UC Santa Barbara, June 11, 2012. *www.news.ucsb.edu.*

Evans, Ben. "And Then Silence: 25 Years since the Rise and Fall of Mars Observer." America Space, September 24, 2017. *www.americaspace.com.*

"Exodus 16." Bible Hub, 2016. *biblehub.com.*

"Exodus 32." Bible Hub, 2016. *biblehub.com.*

Farley, Peter R. "The Anunnaki Branch Grows." Biblioteca Pleyades, 2016. *www.bibliotecapleyades.net.*

Federal Bureau of Investigation (FBI) files on George Adamski, declassified under the terms of the US Freedom of Information Act.

Federal Bureau of Investigation (FBI) files on George Van Tassel, declassified under the terms of the US Freedom of Information Act.

Firestone, Richard B., and William Topping. "Terrestrial Evidence of a Nuclear Catastrophe in Paleoindian Times." *DefendGaia.org,* March 2001. *abob.libs.uga.edu.*

Foulke, Nicole. "The Banyan Trees of Mars." Popular Science, December 17, 2001. *www.popsci.com.*

"Found It! Ice on Mars." NASA Science website, March 3, 2020. *science.nasa.gov.*

Friedman, Janice. "Life on Mars? Scientists Say There May Be 12,000 Olympic-Sized Pools Of Organic Matter on Mars." 2020. *www.ancient-code.com.*

Fuller, John G. *The Interrupted Journey: Two Lost Hours "Aboard a Flying Saucer."* New York: Dial Press, 1966.

Gardner, Lawrence. *Genesis of the Grail Kings*. New York: Bantam Press, 1999.

Gatchell, Bryan. "White Sands Opens its Doors to Trinity Test Site." Fort Bliss Bugle, 2015. *fortblissbugle.com*.

Geoffrey of Monmouth, History of the Kings of Britain. Wikisource, 2019. *en.wikisource.org*.

"George Van Tassel." Federal Bureau of Investigation The Vault website. *vault.fbi.gov*.

Geuss, Megan. "Trinitite: The Radioactive Rock Buried in New Mexico before the Atari Games." Ars Technica website, September 1, 2014. *arstechnica.com*.

"Gilgamesh Tomb Believed Found." BBC, April 29, 2003. *news.bbc.co.uk*.

Gill, N. S. "Mars, Rome's Honored War God." Thought Co., March 8, 2017. *www.thoughtco.com*.

"Giovanni Virginio Schiaparelli." Britannica, 2019. *www.britannica.com*.

Godfrey, Linda. *Hunting the American Werewolf*. Madison, Wisconsin: Trails Books, 2006.

Gray, Richard. "Stonehenge-Style Rocks Spotted on Mars: Bizarre Circular Stone Formation on the Red Planet Resembles the Iconic Pagan Site." *Daily Mail*, September 24, 2015. *www.dailymail.co.uk*.

"The Great Flood: The Epic of Atrahatis." Livius.org, 2015. *www.livius.org*.

"Greg Orme: Biography." Coast to Coast AM, 2019. *www.coasttocoastam.com*.

Greicius, Tony. "Jamming with the 'Spiders' from Mars." NASA, July 13, 2018. *www.nasa.gov*.

——. "Mars Today: Robotic Exploration." NASA, March 7, 2019. *www.nasa.gov*.

Guest, E. A. "The Other Paradigm." *Fate*, April 2005.

Haines, Gerald K. "CIA's Role in the Study of UFOs, 1947–90." CIA.gov, June 27, 2008. *www.cia.gov*.

Haines, Lester. "Assyrian Clay Tablet Points to 'Sodom and Gomorrah' Asteroid." The Register, March 31, 2008. *www.theregister.co.uk*.

Haining, Peter. *The Race for Mars: The Greatest Space Race is On*. London: Comet Books, 1986.

Halton, Mary. "Liquid Water 'Lake' Revealed on Mars." BBC, July 25, 2018. *www.bbc.com*.

Hanks, Micah. "Spaceships on the Moon: Lunar Anomalies, the Phobos Monolith, and Pareidolia." Mysterious Universe, June 14, 2018. *mysteriousuniverse.org*.

——. "The Ancient Nukes Question: Were there WMD's in Prehistoric Times?" Mysterious Universe, December 19, 2011. *mysteriousuniverse.org*.

Hardy, Chris H. *DNA of the Gods: The Anunnaki Creation of Eve and the Alien Battle for Humanity*. Rochester, Vermont: Bear & Company, 2014.

——. *Wars of the Anunnaki: Nuclear Self-Destruction in Ancient Sumer*. Rochester, Vermont: Bear & Company, 2016.

Harris, Paul. "The Australasian Strewn Field." Meteorite Times, 2002. *www.meteorite-times.com*.

"Has the Mystery of the Mars 'Monolith' Been Solved? *Daily Mail*, updated August 6, 2009. *dailymail.co/uk*.

Hass, George. *The Martian Codex: More Reflections from Mars*. Berkeley, California: North Atlantic Books, 2009.

Hertsgaard, Mark. "Mikhail Gorbachev Explains What's Rotten in Russia." *Salon.com*, September 7, 2000. *salon.com*.

Hester, Jessica Leigh. "Everybody Shut Up! We're Listening to Mars." Atlas Obscura, August 3, 2018. *www.atlasobscura.com*.

Hiemestra, Glen. "The Glass Tubes of Mars." Futurist, 2019. *www.futurist.com*.

Hille, Karl. "Mars' Moon Phobos Is Slowly Falling Apart." NASA, November 10, 2015. *www.nasa.gov*.

"HiRISE." NASA, 2019. *mars.nasa.gov*.

"History of Stonehenge." English Heritage, 2019. *www.english-heritage.org.uk*.

Hoagland, Richard. "Preliminary Report of the Independent Mars Investigation Team: New Evidence of Prior Habitation?" Case for Mars II Conference. Boulder, Colorado, July 10–14, 1984.

Hoagland, Richard C., and Mike Bara. *Dark Mission: The Secret History of NASA*. Port Townsend, Washington: 2007.

Hoffman, Russell "Ace." "The Effects of Nuclear Weapons." Animated Sofware, May 10, 1999. *www.animatedsoftware.com*.

Hopkins, Budd. *Missing Time*. New York: Ballantine Books, 1981.

"Human Rights Watch Arms Project." Human Rights Watch website, October 2000. *www.hrw.org*.

"The Immortality Machine." *Ancient Aliens*, History Channel, March 14, 2020.

"In the News: Baalbek. Welcome to the 'Landing Place.'" The Official Website of Zecharia Sitchin, 2006. *www.sitchin.com*.

"Ingo Swann." Info-quest.org. https://*info-quest.org*.

"Iosif Samuilovich Shklovskii. (The Bruce Medalists.)" Sonoma State University Department of Physics & Astronomy website, 2019. *www.phys-astro.sonomaedu*.

"Is This Meteor a Message from Faraway Mars?" *New York World*, November 21, 1897.

"Jack Kirby Museum and Research Center." 2019. *kirbymuseum.org*.

Jessup, Morris K. *The Expanding Case for the UFO*. New York: The Citadel Press, 1957.

Jones, Marie, and Larry Flaxman. "Our Sonic Past: The Role of Sound and Resonance in Ancient Civilizations." In *Lost Civilizations & Secrets of the Past*. Pompton Plains, New Jersey: New Page Books, 2012.

Jordan, Gary. "Can Plants Grow with Mars Soil?" NASA, August 6, 2017. *www.nasa.gov*.

Klimczak, Natalia. "Anubis, Egyptian God of the Dead and the Underworld." Ancient Origins, April 6, 2019. *www.ancient-origins.net*.

Knowles, Christopher. "Mindbomb: John Carter, PKD and the 'Face on Mars' Revisited." Secret Sun blog, August 15, 2012. *secretsun.blogspot.com*.

Kress, Kenneth. "Studies in Intelligence." Central Intelligence Agency, 1977. *www.cia.gov.*

Lamb, Robert. "What Would Nuclear Winter Be Like?" How Stuff Works. *science. howstuffworks.com.*

Lean, Geoffrey. "Sodom and Gomorrah 'Destroyed by a Comet,' Say Astronomers." Independent, March 29, 1997. *www.independent.co.uk.*

Lee, Rhodi. "Did NASA Destroy Europe's Mars Lander? Conspiracy Theory Suggests US Space Agency Shot Down Schiaparelli." *Tech Times,* October 25, 2016. *www.techtimes.com.*

Leonard, George H. *Somebody Else Is on the Moon.* New York: Pocket Books, 1977.

"Little Boy and Fat Man." Atomic Heritage, 2015. *www.atomicheritage.org.*

"Lonar Crater—An Impact Crater." Biblioteca Pleyades, 2015. *www.biblioteca pleyades.net.*

"Lonar Crater, India." Visible Earth, December 15, 2014. *visibleearth.nasa.gov.*

Longnecker, Dwight. "Why Did God Destroy Sodom and Gomorrah?" Patheos, June 25, 2015. *www.patheos.com.*

MacKenzie, Debora. "'Nuclear Winter' May Kill More Than a Nuclear War." New Scientist, March 1, 2007. *www.newscientist.com.*

Mandia, Scott A. "The Little Ice Age in Europe." Suffolk County Community College website, 2015. *sunysuffolk.edu.*

"The Manhattan Project: Making the Atomic Bomb." Atomic Archive, 2015. *www.atomicarchive.com.*

Mann, Charles C. "The Clovis Point and the Discovery of America's First Culture." Smithsonian Magazine, November 2013. *www.smithsonianmag.com.*

"Marconi Testing His Mars Signals." *New York Times,* January 29, 1920. *www .nytimes.com.*

Mark, Joshua J. "Gilgamesh." Ancient History Encyclopedia, March 29, 2018. *www.ancient.eu.*

———. "Nefertiti." Ancient History Encyclopedia, April 14, 2014. *www.ancient.eu.*

Marrs, Jim. *Our Occulted History.* New York: HarperCollins Publishers, 2013.

"Mars Artifacts." *Tripod.com,* 2019. *marsartifacts.tripod.com.*

"Mars Atmosphere Is Supersaturated with Water." *Astrobiology Magazine,* October 2, 2011. *astrobio.net.*

"Mars Exploration, May 22, 1984." Central Intelligence Agency, May 22, 1984. *www.cia.gov.*

"Mars Global Surveyor." NASA, 2019. *mars.nasa.gov.*

"Mars 'Monolith' Fuels Theories of Alien Life." Telegraph, 2019. *www.telegraph .co.uk.*

"Mars Observer." NASA, 2019. *mars.nasa.gov.*

"Martian Moons: Phobos." European Space Agency website, 2019. *sci.esa.int.*

Maynard, James. "Crab-Like Facehugger Spotted on Mars by Curiosity Rover— What Is It?" Tech Times, August 5, 2015. *www.techtimes.com.*

McDaniel, Stanley V., and Monica Rix Paxson (Eds.). *The Case for the Face: Scientists Examine the Evidence for Alien Artifacts on Mars.* Kempton, Illinois: Adventures Unlimited, 1998.

McFall-Johnsen, Morgan and Mosher, Dave. "Elon Musk Says He Plans to Send 1 Million People to Mars by 2050 by Launching 3 Starship Rockets Every Day and Creating 'a Lot of Jobs' on the Red Planet." January 17, 2020. *www.business insider.com.*

"Megabeasts' Sudden Death." *PBS.org,* March 31, 2009. *www.pbs.org.*

Mervine, Evelyn. "Desert Glass." Skeptic Report, June 1, 2005. *www.skepticreport.com.*

"Message Perhaps from Mars." *New York Times,* November 14, 1897.

"Mike Bara Gets it Wrong as Usual." The Emoluments of Mars website, March 24, 2017. *dorkmission.blogspot.com.*

Mills, Ted. "20 New Lines from *The Epic of Gilgamesh* Discovered in Iraq, Adding New Details to the Story." Open Culture, October 5, 2015. *openculture.com.*

Mitchell, Stephen. *Gilgamesh: A New English Version.* New York: Atria Books, 2006.

Mooney, Richard E. *Colony Earth.* New York: Stein and Day, 1974.

"NASA—Report Reveals Likely Causes of Mars Spacecraft Loss." NASA website, April 13, 2007. *nasa.gov.*

Nelson, Bryan. "9 Features on Mars Mistaken for Signs of Alien Life: Banyan Trees." MNN, November 19, 2014. *www.mnn.co.*

"The Nephilim—Giants in the Bible." Beginning and End, May 23, 2011. *www.beginningandend.com.*

"New Evidence Supports Theory of Extraterrestrial Impact." Sacred Geometry International, June 11, 2012. *sacredgeometryinternational.com.*

Nicks, Ron. "Opening a Martian 'Can of Worms . . .'" Enterprise Mission, 2019. *www.enterprisemission.com.*

"Nikola Tesla Articles." Tesla Universe, 2019. *teslauniverse.com.*

Northon, Karen. "NASA Research Suggests Mars Once Had More Water than Earth's Arctic Ocean." NASA, March 5, 2015. *www.nasa.gov.*

"Notes and News." *The Academy, volume 52,* December 11, 1897, page 525.

"Nuclear Winter." Atomicarchive.com. *atomicarchive.com.*

"Nuclear Winter." Originally from *Encyclopedia Britannica. www.britannica.com.*

"Numbers 13:31." Bible Hub. *biblehub.com.*

Oberg, James. "The Dark Side of Space Disaster Theories." Space Review, January 21, 2008. *www.thespacereview.com.*

O'Callaghan, Jonathan. "Ancient Martian Civilization Was Wiped out by Nuclear Bomb-Wielding Aliens—and They Could Attack Earth Next, Claims Physicist." Daily Mail, November 21, 2014. *www.dailymail.co.uk.*

O'Connell, Tony. "Zechariah Sitchin." Atlantipedia website, December 16, 2012. *atlantipedia.ie.*

"Odyssey Finds Water Ice in Abundance Under Mars' Surface." Jet Propulsion Laboratory website, May 28, 2002. *mars.jpl.nasa.gov.*

Orme, Greg. *Why We Must Go to Mars: The King's Valley.* CreateSpace independent publishing platform, 2011.

Palermo, Efrain. "Palermo's Phobos Anomalies." Palermo Project website, 2019. *palermoproject.com.*

Pescovitz, David. "The Mysterious Face on Mars Was First Spotted in 1959." Boing Boing, July 30, 2015. *boingboing.net.*

"Phobos Facts." The Planets website, 2019. *theplanets.org.*

"Phobos: Malfunction or Early 'Star Wars' Incident?" Dark Star 1 website, 2019. *www.darkstar1.co.uk.*

"Physical Evidence of Ancient Atomic Wars Can Be Found World-Wide." MessageToEagle, October 6, 2015. *www.messagetoeagle.com.*

Pippin, Ed. "Tom Corbett Space Cadet." Solar Guard, 2013. *www.solarguard.com.*

"Pre-Hispanic Town of Uxmal." UNESCO, 2019. *whc.unesco.org.*

"Project 1947." Project 1947 website, 2019. *www.project1947.com.*

"Psalm 90." Bible Hub, 2012. *biblehub.com.*

Pye, Lloyd. *Everything You Know Is Wrong.* Madeira Beach, Florida: Adamu Press, 1997.

Pye, Lloyd. "What Is Intervention Theory?" *LloydPye.com,* 2011. *www.lloydpye.com.*

Radford, Benjamin. "Female Figure on Mars Just a Rock." *Space.com,* January 24, 2008. *www.space.com.*

"Radiation Sickness." Mayo Clinic, 2015. *www.mayoclinic.org.*

Ramachandran, Priya. "Lonar—A Walk inside the Meteorite Crater." Happy Feet, October 26, 2013. *happyfeet.us.*

Redd, Nola Taylor. "Phobos: Facts about the Doomed Martian Moon." Space.com, December 8, 2017. *www.space.com.*

Redfern, Nick. *Bloodline of the Gods.* Wayne, New Jersey: New Page Books, 2015.

———. *Contactees: A History of Alien-Human Interaction.* Franklin Lakes, New Jersey: New Page Books, 2010.

———. Interview with Linda Godfrey, April 6, 2003.

———. Interviews with Kimberly Rackley, October 4, 7, 9, and 10, 2019.

———. Interview with Mac Tonnies, July 7, 2009.

———. Interview with Mac Tonnies (via email), March 14, 2004.

———. Interview with Mac Tonnies (via email), September 9, 2006.

———. *Keep Out!* Wayne, New Jersey: New Page Books, 2012.

———. *The Pyramids and the Pentagon: The Government's Top Secret Pursuit of Mystical Relics, Ancient Astronauts, and Lost Civilizations* (Wayne, New Jersey: New Page Books, 2012).

———. "The Sailing Stones of Egypt." Mysterious Universe, October 18, 2012. *mysteriousuniverse.org.*

———. *Science Fiction Secrets: From Government Files and the Paranormal.* San Antonio, Texas: Anomalist Books, 2009.

———. "The 'Top Secret' Brookings Report and Alien Life." Mysterious Universe, January 25, 2019. *mysteriousuniverse.org.*

————. *Weapons of the Gods*. Wayne, New Jersey: New Page Books, 2016.

"Reexamining the Lost Mars Probes of 1989–1993." Keys of Enoch. *keysofenoch.org/*.

"Repairing the Ozone Layer." US Environmental Protection Agency. *yosemite .epa.gov.*

"Robert M. Schoch: About." Robert Schoch website, 2019. *www.robertschoch.com.*

"Robert M. Schoch: Research Highlights—The Great Sphinx." Robert Schoch website, 2019. *www.robertschoch.com.*

Robock, Alan. "Nuclear Winter." The Encyclopedia of Earth, January 6, 2009. *www.eoearth.org.*

Rothman, Lily. "When the U.S. Army Brought Reporters to a Still-Hot Atomic Test Site." Time, July 16, 2015. *time.com.*

Rudd, Steve. "The Epic of Gilgamesh." The Interactive Bible, 2015. *www.noahs-ark.tv.*

Rux, Bruce. *Architects of the Underworld: Unriddling Atlantis, Anomalies on Mars, and the Mystery of the Sphinx*. Berkeley, California.: Frog, Ltd., 1996.

Sangster, Angela. "Who Were the Anunnaki?" TrueGhostTales.com, 2010. *true ghosttales.com.*

Science Daily. "NASA's Curiosity Rover Finds Clues to Chilly Ancient Mars Buried in Rocks." May 19, 2020. *www.sciencedaily.com.*

Sessions, Larry. "What Is Pareidolia?" EarthSky, August 19, 2018. *earthsky.org.*

Sharps, Matthew J. "Percival Lowell and the Canals of Mars." *Skeptical Inquirer* volume 42, number 3, May/June 2018.

Sims, Chris. "How Jack Kirby's Art Helped the CIA Rescue Diplomats in 1979." Comics Alliance, January 5, 2012. *comicsalliance.com/*.

Sitchin, Zecharia. *The Cosmic Code: The Incredible Truth about the Anunnaki who Divulged Cosmic Secrets to Humankind*. New York: Avon Books, 1998.

————. *Divine Encounters: A Guide to Visions, Angels, and Other Emissaries*. New York: Avon Books, 1995.

————. *Genesis Revisited: Is Modern Science Catching Up with Ancient Knowledge?* New York: Avon Books, 1990.

————. "In the News: Baalbek. Welcome to the 'Landing Place'." 2006. *sitchin.com.*

————. *The Lost Realms: Incredible Documentary Evidence of the Extraterrestrial Giants Who Brought Civilization to the New World* New York: Avon Books, 1990.

————. *The 12th Planet*. New York: HarperCollins, 2007.

————. *When Time Began: The First New Age*. New York: Avon Books, 1993.

Smith, Paul H. "What Is Remote Viewing?" International Remote Viewing Association website, 2019. *www.irva.org.*

"Sodom and Gomorrah." *ArkDiscovery.com*, 2015. *www.arkdiscovery.com.*

"Space-Life Report Could Be Shock." *NICAP UFO Investigator*, December 1960/January 1961.

Space.com. "Are Those Trees on Mars?" Fox News, January 13, 2010. *foxnews.com.*

"The Sphinx." History.com. January 4, 2018. *www.history.com.*

Sprague, Ryan. "Third Eye on the Moon." Vocal website. *vocal.media.*

Starr, Steven. "Deadly Climate Change from Nuclear War: A Threat to Human Existence." Nuclear Darkness website, 2015. *www.nucleardarkness.org*.

"The Stealth Catastrophe." *Science Frontiers Online, No. 136.* July–August 2001. *science-frontiers.com*.

Steigerwald, Bill. "Designer Plants on Mars." NASA, 2019. *www.nasa.gov*.

Stieber, Zachary. "Comet Strike Evidence Found in Tutankamun's Brooch and the Sahara Desert." The Epoch Times, October 9, 2013. *www.theepochtimes .com*.

"Stonehenge Bluestones Had Acoustic Properties, Study Shows." BBC, March 3, 2014. *www.bbc.com*.

"The Story of Trinitite." Bradbury Science Museum website. *lanl.gov*.

Strauss, Mark. "Ten Enduring Myths about the U.S. Space Program." Smithsonian Magazine, April 14, 2011. *www.smithsonianmag.com*.

Strieber, Whitley. *Breakthrough: The Next Step.* New York: Harper Spotlight, 1995.

"Sumerian Gods and Goddesses." Crystalinks.com, 2016. *www.crystalinks.com*.

Swann, Ingo. *Penetration: The Question of Extraterrestrial and Human Telepathy.* Plano, Tex.: Swann-Ryder Productions, LLC, 2018.

Swift, Jonathan. *Gulliver's Travels.* New York: Wordsworth Editions Limited, 1997.

"Talk: Selim Hassan." Wikipedia. *wikipedia.org*.

Taylor, Stuart Ross. "The Australasian Tektite Age Paradox." *Meteoritics and Planetary Science*, no. 34, 1999. *onlinelibrary.wiley.com*.

Tellinger, Michael. *African Temples of the Anunnaki.* Rochester, Vermont.: Bear & Company, 2013.

———. *Slave Species of the Gods.* Rochester, Vt.: Bear & Company, 2012.

"10 Things You (Probably) Didn't Know about Stonehenge." History Extra, June 21, 2019. *www.historyextra.com*.

Tikkanen, Amy. "Great Sphinx of Giza." Britannica, 2019. *www.britannica.com*.

Tonnies, Mac. *After the Martian Apocalypse: Extraterrestrial Artifacts and the Case for Mars Exploration.* New York: Paraview Pocket Books, 2004.

———. *The Cryptoterrestrials: A Meditation on Indigenous Humanoids and the Aliens Among Us.* San Antonio, Tex.: Anomalist Books, 2010.

"Traces of Giants Found in Desert." *San Diego Union*, August 5, 1947.

Trager, Martelle W. *National Parks of the Northwest.* New York: Dodd, Mead and Co., 1939.

"Transport of Water Vapor in the Martian Atmosphere." European Space Agency Science and Technology website, September 29, 2011. *sci.esa.int*.

"Twin Peaks in Super Resolution—Left Eye." Jet Propulsion Laboratory website, September 8, 1999. *www.jpl.nasa.gov*.

"UFO." Federal Bureau of Investigation The Vault website. *vault.fbi.gov/UFO*.

Utrecht University. "Reading Martian Rocks in Unparalleled Detail to Find Ancient Water on Mars." May 8, 2020. *scitechdaily.com*.

Velikovsky, Immanuel. *Earth in Upheaval.* New York: Pocket Books, 1977.

———. *Worlds in Collision.* New York: Pocket Books, 1977.

"Volcanic Rock in Mars' Gusev Crater Hints at Past Water." Jet Propulsion Laboratory website, March 5, 2004. *www.jpl.nasa.gov.*

Voltaire. *Micromegas.* The Public Domain Review, 2019. *publicdomainreview.org.*

Von Daniken, Erich. *Chariots of the Gods?* London: Souvenir Press, Ltd., 1970.

Wagner, Stephen. "The Most Mysterious Anomalies of Mars." Live About, May 24, 2019. *www.liveabout.com.*

———. "What's on the Far Side of the Moon?" Live About, May 25, 2017. *www.liveabout.com.*

Wall, Mike. "How the Mars Moon Phobos Got its Grooves." Space.com, November 25, 2018. *space.com.*

Watanabe, Susan (Ed.). "Water: The Molecule of Life." NASA, November 30, 2007. *www.nasa.gov.*

Waugh, Rob. "Crab-Like Alien 'Facehugger' Is Seen Crawling out of a Cave on Mars." Metro, August 5, 2015. *metro.co.uk.*

"Welcome to Meteor Crater." Meteor Crater, 2015. *meteorcrater.com.*

"What Was the Sin of Sodom and Gomorrah?" Got Questions, 2015. *www.gotquestions.org.*

"White Powder of Gold (ORME)." Token Rock. *www.tokenrock.com.*

"Wiggins on the Aerolite." *New York Times,* November 18, 1897.

Wigington, Patti. "Mars, Roman God of War." Learn Religions, December 10, 2018. *www.learnreligions.com.*

Wolchover, Natalie. "'Monolith' Object on Mars? You Could Call it That." Live Science, April 11, 2012. *www.livescience.com.*

"World Speaks to World with Mysterious Signals Through Vast Space—Tesla, the Electrician Says He Received a Message from Mars." *San Francisco Examiner,* January 4, 1901.

Yeo, Amanda. "Watch NASA test the Mars Perseverance rover ahead of launch." May 18, 2020. *https://mashable.com.*

Index

About the Author

Nick Redfern is the author of more than sixty books, including *Keep Out!, For Nobody's Eyes Only, Immortality of the Gods, Top Secret Alien Abduction Files, Men in Black, Bloodline of the Gods, Contactees, The Pyramids and the Pentagon,* and *Weapons of the Gods*. He has been on many television shows, including Travel Channel's *In Search of Monsters*, History Channel's *UnXplained* and *Monster Quest*, SyFy Channel's *Proof Positive*, and National Geographic Channel's *Paranatural*. Nick is a regular guest on *Coast to Coast AM*.

Nick's blog, World of Whatever: *http://nickredfernfortean.blogspot.com*
Nick's Twitter account: *twitter.com/nickredfernufo*
Nick's Facebook page: *www.facebook.com/nick.redfern.73*